Auto Mechanics

Studies in Industry and Society

Philip B. Scranton, Series Editor

Published with the assistance of the Hagley Museum and Library

Related titles in the series:

Mark Aldrich, *Safety First: Technology, Labor, and Business in the Building of American Work Safety, 1870–1939*

John K. Brown, *The Baldwin Locomotive Works, 1831–1915: A Study in American Industrial Practice*

Thomas R. Heinrich, *Ships for the Seven Seas: Philadelphia Shipbuilding in the Age of Industrial Capitalism*

David Hounshell, *From the American System to Mass Production, 1800–1932 The Development of Manufacturing Technology in the United States*

Thomas A. Kinney, *The Carriage Trade: Making Horse-Drawn Vehicles in America*

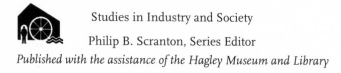

Auto Mechanics

Technology and Expertise in Twentieth-Century America

KEVIN L. BORG

The Johns Hopkins University Press
Baltimore

Printed in the United States of America on acid-free paper
2 4 6 8 9 7 5 3 1

The Johns Hopkins University Press
2715 North Charles Street
Baltimore, Maryland 21218-4363
www.press.jhu.edu

Borg, Kevin L.
Auto mechanics: technology and expertise in twentieth-century America /
Kevin L. Borg.
p. cm. — (Studies in industry and society)
Developed from author's dissertation for the Hagley Program in the History of Industrialization, University of Delaware.
Includes bibliographical references and index.
ISBN-13: 978-0-8018-8606-5 (hardcover : alk. paper)
ISBN-10: 0-8018-8606-6 (hardcover : alk. paper)
1. Automobile repair shops—United States. 2. Automobile industry and trade—Social aspects—United States. 3. Automobile mechanics—United States.
4. Automobiles—United States—Maintenance and repair. I. Title.
TL153.B667 2007
338.4'762928720973—dc22 2006030071

A catalog record for this book is available from the British Library.

CONTENTS

ACKNOWLEDGMENTS

I must break with the usual academic acknowledgments format and recognize up front the significant contributions of my spouse and partner, Jere Borg. Throughout this book's decade-long gestation she has been a patient listener and my most reliable editor. She has read and commented on draft upon draft of every section of every chapter—sacrificing time, emotional energy, and sleep beyond what either of us had anticipated when we began our life journey together. To her I owe a personal and professional debt that I can never repay, though I vow to make the effort for the remainder of my days.

Over the course of many years the research presented here has profited from the help of far too many individuals to recount in detail. Some of them stand out, however, and must be thanked publicly. Ron Tobey's remarkable undergraduate teaching ignited my interest in the history of science and technology and set me on this exciting, if uncertain, path. This book grew out of my graduate studies in the Hagley Program in the History of Industrialization at the University of Delaware. The financial support and intellectual climate of the Hagley Program and the Department of History provided a fertile bed for inquiry and growth. There Arwen Mohun directed my doctoral work and supported my unorthodox research topic from its inception in one of her graduate writing seminars until long after her official responsibilities had been fulfilled. She always allowed me enough freedom to pursue my own ideas and sources with just enough "directing" to maintain a viable research project. In addition, George Basalla, Reed Gieger, Roger Horowitz, Bill Leslie, John Staudenmaier, and Susan Strasser each read my early research, in whole or in part, and offered valuable suggestions, sources, and literature that enriched my thinking, my research, and my writing. Historian Steve McIntyre contributed doubly to this project. First, by plowing a research path into the auto repair industry through his earlier dissertation on the topic and, second, by sharing his research photocopies, his friendship, and many conversations about auto repair research topics. Phil Scranton has been the most helpful and patient editor I could have hoped for. His initial critique of my earliest work on this topic while I was a young graduate student was transformative, and his fertile mind has been a constant inspiration. His sharp editor's pen has urged me toward better writing, and his unfailing support has been crucial to bringing this book to completion. My colleagues at James Madison University—David Ehrenpreis, Fletcher Linder, Mark Thomas, and Steve Reich—have also read parts of my research and helped me sharpen my thinking and presentation.

Financial and institutional support for this project has come from the University of Delaware and the Hagley Museum and Library in the form of a Hagley Fellowship in the History of Industrialization as well as supplemental fellowship and travel funding. A Smithsonian Institution Predoctoral Fellowship supported four months of research in the National Museum of American History's Transportation Collection, Trade Catalog Collection and Archives Center. A Clark Research Grant from the Henry Ford Museum and Greenfield Village allowed me to travel to and use that fine collection. The History Department at University of California, Riverside (UCR), my undergraduate alma mater, granted me Visiting Researcher status while I was researching and writing in California. The Interlibrary Loan staffs of UCR's Rivera Library and James Madison University's Carrier Library have graciously and politely fulfilled my requests for old, hard-bound trade journals, obscure auto repair books, and other diverse resources. My home institution, James Madison University, has supported the dissertation-to-book research through an Edna T. Schaeffer Humanist Award. The Society for the History of Technology (SHOT) contributed a much-needed boost to this book project through the SHOT Writing and Publication Workshop at Woods Hole, Massachusetts, in the summer of 2004. There I gained valuable insights into advanced writing for a general audience from professional writers and editors as well as specific critiques and advice on a portion of this book from Tom Jehn, Larry Cohen, Rosalind Williams and fellow participants Gwen Bingle, Vera Candiani, Maja Fjaestad, Maril Hazlett, Per Hogselius, Anders Houltz, Helen Watkins, Matt Wisnioski, Timothy Wolters, and Shana Worthen. This book, while far from perfect, is immensely better due to revisions I made in response to my many critics and supporters. Despite all of this expert guidance and help, flaws remain and are mine alone.

Auto Mechanics

Technology's Middle Ground

Cars break down. They always have. On a warm spring day in 1901 a man named Robin Damon expected to enjoy the new freedom of automobility—swift individual travel without rails, without schedules, and free of willful horses. Instead, he and his friend spent six hours in the hot sun replacing spark plug gaskets, putting in new ignition points, and replacing a broken battery wire in the friend's stranded "gasoline carriage."[1] The promises of the new technology, it turned out, were conditional. As one of Damon's contemporaries understood it, "The power-driven vehicle is wholly and directly in the care of your head and hand. You are the most independent and absolute monarch locomotion ever produced—until something happens; then no wayside beggar is so poor or helpless."[2]

Today, anyone who depends on a car to get to work or who loads soccer-cleated kids into a minivan or hands a nineteen-year-old daughter an auto club membership card as she heads off to college knows the personal havoc that car troubles can create. Despite media pronouncements that we are living in the "information age," Americans have not yet transcended the automobile age. Although less frequently than in Damon's time, as the twenty-first century unfolds one ubiquitous, troublesome, anxiety-inducing experience persists: cars wear out and break down. They need to be maintained and repaired. This book explores how Americans shaped and defined an occupation responsible for the dirty, difficult, and inescapable task of maintaining and repairing their growing fleet of automobiles.

Much has changed since Damon's frustrating day out, yet car troubles can still make us feel as vulnerable as a wayside beggar. Breakdowns place a sudden, unpredictable check on our liberty, a temporary but certain retraction of the promises of automobility. They betray the myth of the open road and create an anxiety of lost freedom and isolation. Our dependence on others to repair and maintain our cars also links us closely to Damon's experience. Damon complained that he found himself in the hot sun that day because his friend had not performed the necessary maintenance beforehand. American drivers today are far more likely to be totally dependent on others to maintain and repair their cars. Toward the end of the twentieth century *Popular Mechanics* magazine surveyed over three thousand car owners and discovered that nearly 97 percent of them did not perform their own automotive repair work.[3] Modern motorists' anxieties about car troubles are compounded by a much higher reliance upon their cars for daily transportation than Damon could have imagined. Breakdowns and even routine maintenance force us to seek help from others to get to and from work, pick up the kids, or get to the doctor appointment while the car sits in the shop.

Over the last century the automobile industry has struggled, with varying degrees of success, to provide customers with reliable automobiles. Its products have often been marvels of engineering. The earliest automakers experimented with cutting-edge steam, electric, and gasoline technologies to power vehicles incorporating diverse configurations of drive trains, steering mechanisms, brakes, and more. Today's cars bring together the latest composite materials, microelectronics, and satellite uplinks. Despite all that, things still go wrong, parts still wear out, cars still break down—they always have and likely always will. Hence, Americans have come to rely on a specific group of workers to meet this inescapable need.

While uncovering the complex social history of the auto mechanic's occupation, this book also highlights the value of studying technology's middle ground—*middle* not in the sense of being between a higher or lower level of technology or status but as occupying an ambiguous space between production and consumption in which workers maintain and repair artifacts that they neither create nor own. This social space is inhabited by a wide range of people and institutions bringing a variety of interests to bear on the maintenance and repair of technological artifacts and systems.

This middle ground is not limited to automobile repair or to modern consumer societies. Numerous workers in wide-ranging historical and cultural contexts share this relationship to technology, whether one considers the seventeenth-century Dutch tool sharpener depicted in Gerard Ter Borch's painting *The Family of the Stone Grinder* or the nineteenth-century Gypsy tinkers who roamed the English countryside repairing pots, kettles, and household tools. Similarly, eighteenth- and nineteenth-century grain millers hired stone dressers to keep their millstones in good condition, and early modern French papermakers relied on tradesmen with specialized knowledge to repair their waterwheels, mill shafts, and presses. In today's rural Indian villages United Nations International Children's Emergency Fund (UNICEF) development workers have taught lower-caste Indian women to repair hand well pumps, and this technological knowledge has increased the status and social acceptance of these women, even into Brahmin homes. "These were the homes we could never enter," said pump mechanic Savitri Kabirdas. "Our pots could not touch theirs when they filled water. Now they make us sit on the cot and offer us tea and food. They even call us Mechanic Sir." Similar to Americans' reliance on automobiles and mechanics, Indian villagers' dependence upon the well pump vests these women and their technological knowledge with a measure of social power. Wherever such repairs of technology are carried out by a distinct group of workers, these workers and their socioeconomic relationships represent technology's middle ground.[4]

Not all repair work qualifies for inclusion in this middle ground. Women in early-nineteenth-century Massachusetts used their tools and technological knowledge to mend their household clothing, but they were primarily owners of the items under repair and thus lacked the economic distinction between consumer and repairer. Blacksmiths, on the other hand, long hovered at the other end of the spectrum, as they both produced and repaired metal objects, but in the late nineteenth century they shifted more firmly into the middle, when they lost most primary fabrication duties to industrial production.[5]

The twin forces of industrialization and consumerism have certainly changed the contours of this space between production and consumption. Our ability and choice to discard items rather than repair them has made the notion of a tinker coming to the house to repair pots and pans seem quaint and far removed from our own experience. Nonetheless, we are still surrounded by and dependent upon a wide range of workers who repair and maintain our technological systems. Plumbers, airplane mechanics, computer help desk workers, and photocopier technicians all occupy this middle ground and share a relationship with technology which differs in technical and social terms from those of producers, consumers, or users.[6] Further studies of which technologies get repaired rather than discarded, which social groups do the work, and how their technological knowledge both challenges and reinforces their power and status would do much to improve our understanding of one of humankind's unique proclivities. Perhaps more often than we are *Homo faber,* we are *Homo fixer.*

The automobile mechanic's occupation offers a particularly rich opportunity to study *Homo fixer.* Most Americans recognize, if dimly, the central role—for both good and ill—which automobiles have served in twentieth-century American culture. Schoolchildren across the nation, for example, recognize the name of Henry Ford—even if they mistakenly think he invented the automobile. The history of the rise and development of the automobile manufacturing industry has in many ways been the history of American economic growth, decline, rebirth, and ultimately globalization. Yet outside of Detroit, Flint, and other automaking cities, more Americans knew and interacted with automobile mechanics than with assembly line workers, despite the influence of automobile production on the nation's economy and U.S. labor history. Whereas automobile production has been geographically concentrated for much of the twentieth century, auto mechanics and repair shops have existed in virtually every city, suburb, and rural crossroad in America from the first decade until the present. Today almost everyone has a story to tell about this middle ground, whether about their Uncle Frank who was a mechanic in the 1920s or about "D.J." across town, recommended as "the only guy you can trust" to work on your Honda. Inasmuch as automobiles permeated American culture over the course of the twentieth century, so too did automobile mechanics and repair shops.[7]

The technological, social, and economic space in which mechanics operate differs significantly from either the factory building or the driver's seat. The repair shop is where the weaknesses of technology are laid bare; where progress is stalled, repaired, and sent back on the road; where technological failure is the stock-in-trade and the ideal of the well-oiled machine meets the reality of our entropic world. The work processes there resist the predictability, regularity, rationalization, and regimentation so often associated with industrial production and the assembly line. Despite the proliferation of mass production techniques in the auto industry, each of the cars rolling out of the factories encountered unique drivers, diverse road and climate conditions, and differing rates of rust, wear, and abuse. Putting such complicated technology into wide use meant that each failure, each regular service, each repair, had the potential to become frustratingly unique. Thus, a constant tension simmered in the auto repair industry between the desire of automakers and dealers to standardize repairs in the manner familiar to production and the particularity of failures that required skilled, knowledgeable workers to repair them.

Few occupations conjure stronger class and gender associations than that of the auto mechanic. By the middle of the twentieth century it had become a partially stigmatized occupation.[8] Most today see it as a dirty, blue-collar, working-class, male job. Yet the mechanic's position carries social ambiguities. Asked to take over the automotive review column at a general readership magazine, journalist Lesley Hazleton decided to learn more about cars by working as a mechanic's apprentice in a small Vermont shop in the late 1980s. She recognized at the outset that she was committing a "class transgression" in the eyes of her friends by descending from journalism to manual labor, even if only for a summer. Even a homeless man in the park gave her a verbal lashing over her seemingly irrational choice. She was also embarking on a gender transgression, for despite other occupational inroads secured by late-twentieth-century feminism, in 1990 female mechanics made up less than 2 percent of the occupation. It would be safe to wager that few reading this book would embrace their own child, spouse, or loved one making a permanent career move similar to Hazelton's brief experiment. Many motorists like and trust their mechanic—whom they refer to as "my mechanic"—but few would be willing to extend that respect to the occupation generally.[9]

Nonetheless, the auto mechanic still wields a special power and influence, something Hazelton discovered unexpectedly during her apprenticeship: "What I could not grasp until I saw it," she wrote, "is that the moment something goes wrong with the car, the whole balance of power changes. For the exact span of time that a car has problems, the owner sees the mechanic as possessed of some mythic power." She recounted the time a doctor brought his BMW 535 in for a new exhaust system. Placed in the awkward position of having to wait in the shop while the mechanics worked on his car, the doctor tried to ingratiate himself by telling off-color jokes and complaining that he really did not make that much money, perhaps no more than they did. Yet once the car was finished the doctor flipped the mechanic a twenty-dollar tip, hopped into his repaired status symbol, and sped away. Hazleton observed, "The moment the mechanic fulfills his role, he loses his aura, his mystique, his magical powers, and becomes once again a working-class man to be paid off and forgotten—until the next time."[10]

This book seeks to understand the origins and evolution of the automobile mechanic's social identity and the nature and extent of the "mythic power" that Hazelton glimpsed that summer. It asks whether Americans think poorly of auto mechanics because of the kind of dirty work they do, or if they devalue mechanics' work because they believe auto mechanics to be socially, morally, or intellectually beneath them. It asks how and why so many motorists experience such powerless dependency when interacting with mechanics. Moreover, in a society that exalts and celebrates technology in general and the automobile in particular, why does the mechanic's technological expertise not further elevate the practitioner's social status? How can we understand the occupation's ambiguous social status—between stigmatized, dirty service work and technological expertise—as both servants and savants? This study also seeks to discover how an occupation rife with social tension sustained itself. Where did recruits come from? Who chose to become auto mechanics at various times and why? What attractions and benefits did they see in repairing cars? Who was encouraged to pursue these attractions and who was not? How did entrants acquire the skills, knowledge, expertise, and ultimately mythic power of the automobile mechanic, and what social powers acted with or against them as they pursued their goals?

The following history traces a century-long story punctuated at each end by epochal technological change: at the beginning by the introduction of the automobile and at its conclusion by the proliferation of the computer and its incorporation into automotive mechanical systems. Between those bounds lies a history that is as much about maintaining social boundaries, hierarchies, and relationships as it is about maintaining automobiles. It is not a romanticized story of "authentic" mechanics or of a golden age when cars were simpler and all mechanics were good. Auto mechanics' status, knowledge, and trustworthiness have always been contested. As Americans adopted early automobiles, the mechanical demands of the new technology presented a range of actors with opportunities either to replicate or to undermine existing social hierarchies and relationships.

A period of rapid technological change and promising potential contributed to social flux and considerable anxiety as the old sociotechnical hierarchies that had stabilized around late-nineteenth-century horse-drawn transportation splintered and all but collapsed. For the first three decades of the twentieth century Americans in different settings and situations mobilized an astonishing array of resources in their efforts to nudge the emerging automobile-centered structure toward their own favor. Wealthy motorists, chauffeurs, bicycle mechanics, blacksmiths, educators, legislators, automakers, and even the U.S. Army concerned themselves in some way with the question of who should repair our cars. The history of this process unfolds in the first three chapters and illustrates the complex ways in which technology and society coevolve, how they recursively influence each other. In this account mechanics' relationship to the automobile—a technology strongly associated with freedom, power, and progress—both challenged and reinforced evolving American hierarchies of gender, class, and race.

By the late 1930s a new, relatively stable sociotechnical structure formed around mass automobility, symbolically and functionally institutionalized in public education and high school "auto shop." The story in chapter 4 thus becomes one of job entrants "learning about" and "learning to be": learning about the mechanics of automobiles and learning to be auto mechanics; how to repair worn-out clutches and how to view themselves in relation to others.[11] This history uncovers how mechanics, and also those who interacted with them, developed and perpetuated expectations of who auto mechanics should be, what their technological and social roles should be. Not all mechanics, nor even necessarily good mechanics, came out of the public school system. Rather, automobile vocational education helped large numbers of young Americans decide whether to become mechanics and aided those who chose to do so in forming their social identity. Vocational education narrowed, ossified, and reinforced the class, race, and gender disparity between mechanics and motorists by locating crucial technological knowledge within a narrow demographic band. Some in the industry still tinkered with these emerging sociotechnical hierarchies, as we will see in chapter 5, but by the outbreak of World War II the new equilibrium had gained considerable power and inertia.

Readers will find some actors less prevalent in this early story than might be expected. Automakers invested varying degrees of effort into mechanic training during the first half of the twentieth century, but they aimed their formal training efforts primarily at mechanics already working in dealerships. Instead, automobile dealers and independent shop owners experienced the urgent need for mechanics more directly than automakers and consequently became more important actors in the coalition of interests that supported vocational auto shop from the 1920s forward. Gas stations and oil companies also contributed less influence in the early years than might be expected, remaining largely associated with very light maintenance work, or "TBA" ("tires, batteries and accessories"), as the petroleum industry called it. The notion of the "super service station," which offered extensive auto repair work under the banner of a major oil company, did not emerge until the structures for creating and defining the mechanic's occupation were well established.[12]

World War II presented only fleeting challenges to this sociotechnical system, but as chapter 6 reveals, the postwar decades laid bare the paradox inherent in a culture that celebrated automobiles at the same time that it devalued related occupations in technology's middle ground. As the number of cars and roads in America reached new levels in the postwar decades, so too did Americans' dependence on automobiles for daily transportation. By 1963, 82 percent of Americans who traveled to work relied on their cars to do so. Meanwhile, an emerging car culture media popularized under-hood tinkering, celebrated hot-rodding, and made household names of race car drivers and their mechanics. As the 1950s turned into the 1960s and 1970s, however, mechanics found their work increasingly complicated and less rewarding, both financially and socially, and their numbers thinned relative to automobile ownership. Plenty of boys poked around under the hood of their own cars, but fewer and fewer wanted to repair other people's cars for a living. As a group, mechanics remained organizationally weak, and the sociotechnical system that stabilized around their occupation at mid-century left them vulnerable to many of the anxieties and changes ensuing from auto dependency. They soon became the focus of an unprecedented wave of consumer distrust and suspicion.

By the late 1960s Americans had gradually transformed automobility into auto dependency, and like a dependence on drugs, it began to wrack the body politic with consequences. After more than a decade of orgiastic enthusiasm for tail fins and Motoramas, increasing numbers of Americans began to recognize the social and environmental costs of mass automobility. Automobile fatalities and injuries mounted each year, while tailpipe emissions blanketed playgrounds with lead deposits and filled metropolitan skies with irritating haze. Each investigation into these issues implicated automobile mechanics as part of the problem, compounding the long-standing mistrust of the occupation and pushing politicians, industry groups, and consumer watchdog organizations to investigate the auto repair industry repeatedly from the late 1960s through the 1990s. Chapter 7 explores this shifting political climate and how the rules and regulations that followed turned the auto mechanic's occupation into a very closely monitored occupation by the late twentieth century. This distrust and scrutiny became manifest in automotive technology and by the end of the century had contributed to the forces that turned automobiles into rolling computers. The embedding of dozens of electronic sensors and controls in the mechanical systems of automobiles and the routing of engine control functions through a central computer resulted in part from the social stereotype of mechanics. At the same time, it fundamentally challenged the way mechanics understood and worked on cars and presented the greatest challenge to the mechanic's sociotechnical identity in eighty years.

Thus, across its chapters, this story also explores what auto mechanics know and how they know it, which differs from that of engineers and distinguishes mechanics' middle ground position between producers and consumers.[13] A key skill among auto mechanics remains the ability to diagnose the cause of technological failure or malfunction, and for most of the twentieth century such diagnostic skills depended upon what can best be described as a visceral knowledge of automotive technology. Until very recently, much of what auto mechanics knew came unmediated through their five senses—the sound of pre-ignition, the smell of gas-fouled plugs, the pattern of wear visible on individual parts. The tools they used were extensions of their physical body. Wrenches and screwdrivers added leverage, feeler gauges heightened manual sensitivity, and stethoscopes pinpointed sounds. Even when diagnosing electrical ignition system problems, mechanics often relied on visual qualities such as the color of the spark or how large a gap it could or could not jump. Some have called this aspect of an automobile mechanic's skill "kinesthetic," or bodily, knowledge. Such experience-derived knowledge forms the basis of what we call "skill." This type of sense-based, or "tacit," knowledge is difficult to describe verbally or to communicate in writing and therefore has not been easily codified, captured, transcribed, or transmitted,[14] despite the numerous attempts we will see by automakers and employers to do all of these things. On the one hand, this has made it difficult for employers to "deskill" mechanics' work, as has occurred in some production settings.[15] On the other hand, it has made their mechanical expertise difficult to abstract and formalize, preventing the formation of a professional status around their expert service.[16]

We can better understand the historical and social significance of the auto mechanic's visceral knowledge by contrasting it with the electrician's knowledge in the early twentieth century. Direct visceral knowledge of electricity could be deadly. Even in relatively safe, low-voltage, low-amp circuits, electricians could not directly sense the flow of electricity. They could not "feel" resistance. Significantly, the central item in their toolbox was not called a tool but an "instrument": a voltmeter, likely joined by an ohmmeter. Both instruments carried with them the aura of the scientific laboratory from whence they came. They translated electrical properties into abstract numerical representations, which could then be placed into mathematical equations for diagnosing, predicting, and manipulating the behavior of a circuit. This abstract, representational knowledge of electricity as mediated numbers was fundamental to understanding electricity-based technologies such as early radio.

In the first half of the twentieth century the distinction between mechanical and electrical epistemology, or ways of knowing a technological subject, formed the basis for a growing social divergence between automotive and electrical knowledge. A subtle hierarchy became institutionalized in public education, mirroring Western society's long tradition of associating abstract "head work" with social privilege and tactile manual work with the lower classes. Auto shop courses rarely included more than scant math or science components, whereas the curricula for electricity-based vocational courses included more science and considerably more mathematical work.[17] Rarely did general auto mechanics feel themselves well prepared to tackle difficult electrical problems. Instead, a subspecialty of automotive electrical repair businesses emerged by the 1920s and persisted through most of the twentieth century.

In the late twentieth century computerization struck at the epistemological and social core of the auto mechanic's practice. Automobiles of the 1990s were not just "more sophisticated"—each generation of automobiles has been sophisticated relative to its time. Rather, increasingly electronic automobiles called forth different diagnostic skills and more analytic ways of knowing the affected systems and malfunctions encountered in the shop. In the last ten years computer chips and electronic sensors have come to dominate the mechanical functions of automobiles. Consequently, modern mechanics now need to be equally adept at analyzing digital readouts and ferreting out bad sensor data as they are at listening for knocks.

Automobile repair shops need no longer be bastions of blue-collar strength and physical skill. Modern garages are filled with sophisticated computerized diagnostic instruments, but young men and women with the analytical abilities and training required to work with those systems might also work in white-collar settings. They are deterred from entering the auto mechanic's trade by social and cultural cues a century in the making which encourage them to choose other, higher-status careers. The "technization" of the auto mechanic's work, as some call this process,[18] is opening the formerly closed and stable sociotechnical ensemble as educators, employers, and industry leaders try to rethink and reshape the occupation to make it more attractive. The results of this epochal technological change have not yet played out. This is not the end of the story of technology's middle ground, though it is perhaps the end of an era. We have indeed entered the information age and find it now blending with our continuing automobile age. The next generation of automotive technology—whether internal combustion, hybrid, or hydrogen—will incorporate yet more sophisticated electronics, but it will still wear out and break down. How we value the knowledge and practice of those who take up the service and repair of these vehicles has entered a period of potential negotiation; with attention the profession might become attractive again to a wide range of recruits.

I bring more than academic interests to the study of American auto mechanics, and my personal experience has surely influenced my understanding of the history I tell. I grew up around cars and was determined to be a mechanic from as early as I can remember. It was in my blood, it seemed, or at least deeply seated in my psyche. In 1960 my parents relocated from rural Nebraska to southern California, where my father opened a tire store and struggled for years to stay afloat in a very competitive business. Born a year later, I spent much of my childhood around cars, tools, and mechanics. I loved tinkering with things mechanical and tactile—I still do. My mother likes to tell a story that predates my own memory. I must have been about four years old. My three older siblings were all in school, and my mother needed to go into the tire store, for which she did much of the bookkeeping. As a matter of routine, she took me along. Through the large glass window in the office I could see the shop and discovered the alignment pit. For years to come my siblings and I found this concrete-lined hole in the shop floor nearly irresistible. It had so many attractions. Subterranean and fort-like, it contained some of the fanciest, brightest, most sophisticated looking equipment in the shop with lights, mirrors, bubbles, and dials. Interesting in retrospect, some of this equipment displayed a maker's logo that looked like a child's teddy bear. "The Pit," as we called it, had "planks" to walk and turntables on which you could do the twist. The adults told us, of course, that we were not supposed to play in the Pit, which simply magnified our interest in this fascinating space. On the day my mother recalls, I watched intently from behind the protected office space as my Uncle Jim worked on cars in the alignment pit. When we returned home, she remembers walking out into our garage and discovering me on my back, under the front end of my red wagon, mimicking what I had seen in the Pit.

My instinct for and interest in how things worked were thus fed by my early and constant exposure to the working end of cars at the tire store. My two brothers, my sister, and I spent endless summers playing among the material world of mid-twentieth-century American car culture. We explored and romped among the half-dozen old cars my father kept stored behind the shop. My sister and I learned to drive a stick shift at about ten or twelve years old by running the little Renault to and fro on the small lot behind the shop. All of us played among the old engines, springs, shock absorbers, hubcaps, and brake drums that would pile up until the scrap iron man came to haul them off. We clambered about the stock car Dad sponsored at the local speedway and rooted for that car every Saturday night. As each of us boys grew strong enough to be useful, we swept the shop, cleaned customers' cars, "busted beads" on split-rim truck tires, balanced and rotated tires, and became quite good at handling the pneumatic lug wrench—imagining ourselves in training for a NASCAR pit crew. It is clear to me now that I also embraced cars in part as a failed attempt to connect with a distant and distracted father. Mastering tools and cars was a way for a slightly built and mildly bookish boy to get noticed in a man's world.

At school I excelled in vocational courses but put minimal effort into academic subjects. During my junior high and high school years I took woodshop, metal shop, mechanical drawing, carpentry, and welding. My high school did not have an auto shop, so I enrolled in the community college auto shop for high school course credit. Students were not formally "tracked" in those years, but my choice to eschew academic subjects followed a well-marked path and went unchallenged by my family, my school guidance counselors, and my teachers. What I was doing was entirely unremarkable for a teenage boy from what by then was a "broken home" and a family of modest means.[19]

I landed my first high school job outside of the tire store at a local full-service Shell station, where I eventually took over the Saturday mechanic duties: oil changes, tune-ups, tires, belts, batteries, hoses, and the like. I continued to work in auto repair–related jobs after high school for a couple of years until a number of factors converged to alter my life's trajectory. Cars were growing more complex and my "commonsense" knowledge of mechanics seemed less and less sufficient. My emotional and spiritual maturity had reached the point that I could safely question my motives and goals for becoming a mechanic. As important, my mother allowed me to return home to begin attending the community college. I initially thought I would learn what I needed in order to be a better mechanic. It turned out that I enjoyed the academic courses more than I had anticipated. With new inspiration I made a clean break from my grease monkey roots and spent the next decade of my life getting married, working, and pursuing a bachelors' degree in history. It was not until halfway through my graduate studies in the history of technology that I looked back at my own experience with automotive technology and at the century of American history out of which it emerged.

I reveal these things about myself to let the reader know that I bring deep personal interest and experience to this study. In the pages that follow I rarely make this point of view explicit, but it has surely influenced my understanding of the history I tell. I bring what I hope is a synthesis of lived experience and disciplined inquiry to the history of the auto mechanic's occupation and technology's middle ground. We live in a world of sophisticated technologies and complex social hierarchies. It behooves us to know how they are interlaced, how they got that way, and how they might be different.

CHAPTER I

The Problem with Chauffeur-Mechanics

On a summer Sunday in 1906 a *New York Times* headline told a paradoxical story: "Chauffeurs Lord It over Their Employers."[1] Chauffeurs became a serious problem for wealthy motorists during the first decade of the twentieth century. They extorted commissions and kickbacks from garage owners, took their employers' cars out for joyrides at all hours, and exhibited a brazen disregard for social decorum. They did not behave as servants. Between 1903 and 1912 howls of protest arose over chauffeurs' arrogance and insubordination, and the pages of the automotive trade press overflowed with letters, articles, and editorials describing, complaining about, and offering solutions to the "chauffeur problem." As the first group of workers with the primary duty of caring for the automobile, these early chauffeurs represent the first automobile mechanics, and their employers' chauffeur problem marked the first sign of trouble for the new occupation. The habits and attitudes that wealthy Americans brought to their use of automobiles, the sensibilities of those they employed, and the requirements of automotive technology all combined to produce the first major struggle to define the proper social relationships between those who owned automobiles and those who cared for them.

The invention and commercialization of automobile technology created the automobile mechanic's occupation de novo. That is, there certainly were no auto mechanics before there were autos. Yet studies in the history of technology have shown clearly that invention and innovation do not occur in a vacuum. New ideas evolve from and are expressed by analogical and metaphorical connection with familiar ideas.[2] The early care and maintenance of automobiles reveals a similar phenomenon. Important historical connections to earlier occupations and relationships colored the earliest interactions between motorists and mechanics. These preexisting contexts affected the interactions between mechanics and motorists at the work site, whether in an urban garage, a country inn, or a blacksmith's shop, and helped establish the basic shape of the industry and the social outlines of the mechanic's occupation.

The Roots of the Chauffeur Problem

The roots of the chauffeur problem lie in the attitudes that wealthy and upper-middle-class Americans held toward personal transportation in the late nineteenth century, before the adoption of the automobile. Late-nineteenth-century American society relied on horse-drawn carriages, coaches, and carts for personal conveyance.[3] Such technology required constant attention and care both for the horse and for the vehicle. Wealthy Americans almost always placed the care and maintenance of their horses and carriages in the hands of a hired "coachman," who acted as a superintendent of transportation.[4] Employers expected him not only to be a skillful driver but also "to keep his charges in condition, his equipages, equipments, stable, etc., in order, and himself and his subordinates presentable and up to their duties." He supervised the actions of underlings such as grooms and stable hands; took responsibility for the ordering and inspection of fodder and bedding; supervised the feeding, dieting, blanketing, shoeing, and grooming of the horses; inspected and maintained the condition of the carriages and harness; and drove whenever the owner or his wife used the carriage.[5]

Coachmen enjoyed a slightly higher status and greater independence than their indoor, or "house" servant, counterparts. They often had living quarters in or near the stables and received a salary in return for their services. Some employers even granted their coachmen a bit of latitude in the fiscal operation of the stable, allowing them to pocket commissions from stable suppliers. This practice was not without precedent: in the nineteenth century some merchants and grocers, in order to attract business, paid cash or in-kind commissions of 5 to 15 percent to domestic servants of wealthy households who did their employers' shopping. One contemporary advised employers: "If he [a coachman] shoulders a little on the wages—i.e., hires his men for a bit less than you pay, overlook it so long as service is satisfactory—there are bound to be perquisites in all trades, and, if successful, you have had some pickings yourself in your own business."[6]

Despite such privileges, a clear master-servant hierarchy formed the basis of the social relationship between gentleman masters and their coachmen. Beyond the formal duties wealthy Americans had specific expectations regarding the character and behavior of their coachmen. According to one turn-of-the-century gentleman's manual, "A [coachman's] position . . . requires experience, judgment, honesty, sobriety, method and tact, a combination of faculties and acquirements not frequently met with even in much higher walks of life."[7] Gentleman masters knew that their coachmen were not of the "higher walk of life," so they believed it was important that their servants display the proper deference. Wealthy employers expected their coachmen to wear livery—a clear outward sign of servile status—and they expected them to obey orders (see fig. 1).[8] Gentlemen were advised: "When a master gives an order, the coachman or groom should touch his cap, reply, 'Yes sir,' or, 'Very good, sir,' and obey without further comment, unless there is some very good reason for him to speak. . . . Obedience and discipline are necessary and your man, or men, should, if ordered, put a horse in backward without any question."[9] We cannot assume that the gentlemen-masters could always elicit or enforce such deference from their coachmen, but for much of the nineteenth century coachmen fulfilled their employers' expectations often enough that a relatively stable social structure developed around elite horse-drawn transportation.

After 1900, as wealthy Americans began purchasing automobiles, they chose large, expensive, gasoline- and steam-powered touring cars. Early automobile types ranged from small, inexpensive one- and two-cylinder tube-frame buggies and high-wheeled wooden wagons to heavy, expensive electric vehicles and four-cylinder gasoline- and steam-powered touring cars. Wealthy motorists with the means to employ coachmen and chauffeurs purchased almost exclusively large, expensive, gasoline- and steam-powered touring cars such as those offered by Panhard-Lavosier from France; Napier from England; and Packard, Pierce, White, and Stevens-Duryea in the United States (see fig. 2).[10] Adopting the automobile into the existing elite transportation structure, however, meant accommodating new mechanical requirements. Driving the first automobiles proved challenging: simply getting from point A to point B without event was rare. Early users battled easily punctured tires, broken axles and springs, broken drive mechanisms, and myriad other problems.

In the absence of corner gas stations and auto parts chain stores, motorists had to prepare themselves for inevitable breakdowns by carrying ample mechanical provisions with them on trips. Articles in *Horseless Age* recommended bailing wire and a ball of twine as standard motoring accessories. Andrew Lee Dyke, in his 1903 book, *Diseases of the Gasoline Automobile and How to Cure Them,* listed over twenty-five items—ranging from wrenches and pliers to hammers, cold chisels, and asbestos gasket material—which should be included in a touring car's tool box (see fig. 3).[11] As one enthusiast put it, to become a "complete master of the art of driving a self propelled vehicle . . . you must, in the first place, be a good mechanic."[12] Motorists who were not good mechanics found themselves having to negotiate for the services of the farmer, the blacksmith, or the village mechanic in order to get themselves and their machines home again.

Early automobiles also required much more intensive routine maintenance than later models. During the motoring season—generally from spring through fall—the oil in most gasoline-powered vehicles had to be changed nearly every week. Valve seats had to be reground several times during the season and occasionally required regrinding while on the road. The 1907 Pierce-Arrow came from the factory with an under-seat tool kit containing extra intake and exhaust valves.[13] Carbon buildup caused by inefficient combustion of poor-quality gasoline had to be scraped periodically from spark plugs, combustion chambers, and piston crowns. Problems also plagued steam-powered touring cars, most stemming from mineral-laden water and impure fuel. No touring class vehicle escaped a fairly intensive maintenance schedule.[14] One writer for the juvenile market wrote humorously of those who "scud along in their automobiles at twenty miles an hour with the whole family around them": "Seemed so their morals grew fat an' flabby an' shif'less. . . . More 'n half of 'em give up church an' went off on the country roads every Sunday. All along the pike from Pointview to Jerusalem Corners ye could see where they'd laid humbly on their backs in the dust, prayin' to a new god an' tryin' to soften his heart with oil or open the gates o' mercy with a monkey wrench."[15]

To be sure, some wealthy motorists relished working on their own cars. They saw such mechanical interaction with the new technology as the object and challenge of motoring, and the more youthful among them sensed the sexual privilege promised by automotive prowess. Nevertheless, most wealthy motorists simply wanted to enjoy the exhilaration of speed and the freedom of long-distance travel without rails which automobiles offered. More than that, they wanted to share these experiences with their social peers. These motorists viewed the automobile trip as a social setting within which the mechanical demands became a distraction, a nuisance, and possibly even an embarrassment. Therefore, rather than care for the vehicles themselves, a significant number of wealthy motorists sought to employ their own chauffeurs, servants who would perform the duties of driver and of mechanic: a "coachman" for their automobile.

An alternative for these motorists would have been to have their vehicles maintained by one of the independent auto repair businesses that were sprouting up in blacksmith's shops, bicycle shops, and livery stables. Yet such businesses remained uncommon in the first decade of the twentieth century, so the chances of breaking down far from any repair shop were great. Moreover, general mechanics' and blacksmiths' shops often took too long to discover the source of a problem. "The automobile is such a new thing," wrote one Massachusetts motorist in 1901, "that almost every machinist is anxious to work on one, and while at the job he tries to find out all he can about the machinery."[16] The inconvenience and uncertain cost of these exploratory repair practices repelled many wealthy motorists. Instead, they expected that a private chauffeur-mechanic would always be available and would know the particular needs of his employer's vehicle very well, keeping repair expenses and inconveniences to a minimum.

Thus, rather than care for the vehicle themselves or rely on independent repair shops, a significant number of wealthy motorists hired chauffeurs and gave them the responsibility for the vehicle's full-time care and maintenance, much as they had done with their earlier coachman. The chauffeur's employer typically expected that he, like the coachman before him, would live on the property with the automobile, receiving board and lodging in addition to his pay. If lodging was not convenient, as in large cities, employers expected a chauffeur to find his own lodgings and increased his pay accordingly; in such a case the chauffeur would likely spend his day at a city garage where the car was stored, awaiting orders from his employer. Either way, the chauffeur attended to the maintenance and repair of the automobile, at times including extensive fabrication or mending of broken parts—skills that all early auto mechanics needed to have, given the undeveloped parts supply system of the time (see fig. 4). A chauffeur might, in addition, supervise a wash boy, just as the coachman had done for the groom and the stable hand. Wealthy motorists also expected their chauffeurs to wear servant's livery, which one fashion observer described as "cut very nearly in the same manner as the coachman's great coat," and to drive whenever the owner or his wife used the car (see fig. 5). Above all, they expected their chauffeur to be deferential and obedient: "after all he is a hired man and should keep his place."[17]

Wealthy motorists clearly desired that their coachmen make the transition to chauffeur, because coachmen already knew their place. In the words of one editorial the coachman "is better fitted for the work [of a chauffeur] than any other class of operatives, being imbued, first of all, with some measure of comprehension of his social position and, second, with fidelity."[18] A retrained coachman, in other words, would be more likely to continue exhibiting the deferential social behavior familiar to their employers than would, say, a machinist. With the help of training programs established by manufacturers such as Locomobile and Pierce-Arrow and by the Boston and New York branches of the Young Men's Christian Association (YMCA), some coachmen successfully made the transition. New York society notables Andrew Carnegie, Mrs. Russell Sage, and E. D. Morgan all had their coachmen retrained as chauffeurs at either New York's West

Side YMCA or the New York School of Automobile Engineers on West Fifty-sixth Street. A Locomobile representative observed, however, that the best students at the company's school for chauffeurs had been "recruited from the ranks of machinists, repairmen, or assemblers. . . . It is, of course, possible to make a good chauffeur out of a man who has not had mechanical training, but it takes very much more time to do it."[19]

If retraining a trusted and deferential coachman was the ideal solution to the chauffeur problem, it was only rarely accomplished. Whatever analogy employers might have drawn between the coachman and the chauffeur, the two jobs required very different kinds of knowledge. The knowledge and skills of animal husbandry did not necessarily translate into mechanical ability. As one editor wrote, "The only attainments that a coachman, per se, has for the position of chauffeur are his familiarity with what are called the rules of the road and his knowledge of the social relations of the driver of a vehicle to his employer. . . . the man who has a liking for the position of coachman is not likely to have taste or talent for mechanical work."[20] Little evidence has emerged to substantiate the occasional claim that wealthy car owners simply instructed their coachmen to move from looking after their horses and carriages to taking care of their automobiles. While none have studied the rate of transition from coachman to chauffeur in the United States, Nicholas Papayanis notes in his study of Parisian coachmen that horse cabbies, especially the older ones, did not often make the transition to "motor conductor." He further notes that the large cab company Compagnie Générale des Voitures à Pàris (CGV) experienced difficulty in maintaining its policy, set forth in 1900, "that the personnel of [motor] conductors shall be taken exclusively from among [its] coachmen." Despite its best efforts, the CGV found it could not easily make competent mechanics of its coachmen.[21]

Despite being unable to recruit their chauffeurs exclusively, or even largely, from the ranks of trusted coachmen, wealthy motorists still attempted to replicate within their garages the master-servant relationship that they had known in their stables. They did not take into account the real scarcity of experience with automobiles which existed in society at large. When staffing their horse stables, wealthy gentlemen could draw on an array of workers who embodied centuries of practical knowledge about horses and horse-drawn transportation—a technology with which gentlemen themselves were very familiar. Such was not the case with automobiles. Wealthy motorists' expectations that they could leave their chauffeurs in charge of their cars, while knowing almost nothing about the machines themselves, gave chauffeurs the freedom and resources to challenge their im-

TABLE 1.
Age Profile of Chauffeurs Compared to All Employed Males in Major U.S. Cities, 1910
(Percent)

	16 to 20 years old	21 to 44 years old	45+ years old
Boston:			
Chauffeurs	9.5	85.1	5.4
Male Workers	9.9	63.0	25.7
Chicago:			
Chauffeurs	16.0	79.1	4.8
Male Workers	11.7	64.5	22.1
Los Angeles:			
Chauffeurs	20.5	76.3	2.5
Male Workers	8.3	63.8	21.5
New York:			
Chauffeurs	10.9	85.0	4.0
Male Workers	12.1	64.7	21.9
Philadelphia:			
Chauffeurs	12.6	82.9	4.4
Male Workers	11.4	60.8	25.5
San Francisco:			
Chauffeurs	12.5	85.2	2.3
Male Workers	8.1	68.0	23.0
St. Louis:			
Chauffeurs	16.2	78.8	4.6
Male Workers	11.5	62.6	23.8

SOURCE: Based on data drawn from U.S., *Thirteenth Census of the United States,* vol. 4: *1910,* 152–53, 181, 540, 545, 572, 575–78, 589.

NOTE: Percentages do not total 100 due to rounding and the exclusion of workers under the age of sixteen.

posed servant status. Chauffeurs used their knowledge of the new technology to stake out as much material and social space for themselves as possible.

Chauffeurs have left scant historical or documentary evidence of themselves, but a close examination of the census data can provide some insights about their identity. In 1910, the first year the U.S. Bureau of the Census listed chauffeurs as an occupation, 45,785 chauffeurs worked nationwide.[22] Seven hundred and eighty of them were factory chauffeurs who demonstrated cars for customers or otherwise drove for the manufacturers. Of the remaining 45,005 chauffeurs, most were concentrated in the urban centers of the Northeast and Mid-Atlantic states.[23] New York City accounted for 9,355 chauffeurs, or about 6 chauffeurs per 1,000 employed males. Boston motorists employed 1,285 chauffeurs, or about 13 chauffeurs per 1,000 employed males. Philadelphia chauffeurs numbered 1,806, or about 3.5 per 1,000 employed males. In the West, San Francisco had 642 chauffeurs, Los Angeles 590, and Denver 215. St. Louis, with 754 chauffeurs, was the southernmost major city for which the occupation was listed (table 1).[24]

Nationwide, only thirty-three were female, and regardless of geographic region, chauffeurs were younger than the local male working population. In Boston, for example, where 26 percent of all employed males in 1910 were listed as forty-five years of age or older, only 5 percent of employed chauffeurs were listed in that age group. Instead, 85 percent of the chauffeurs were listed as being between the ages of twenty-one and forty-four, while only 63 percent of all male employees were between those ages. This generally younger age profile of chauffeurs was consistent for every major city that listed the occupation as well as for the nation as a whole.

Chauffeurs were also more likely to be native-born whites than the local population of male workers or, to a lesser degree, to be black. They tended not to be first-generation immigrants. Again looking at Boston, where census workers classified nearly 48 percent of the male workforce as foreign-born white, less than 33 percent of the chauffeurs were so classified. Rather, 61 percent of the chauffeurs were native-born white, versus just over 49 percent of all male workers. Almost 6 percent were black, versus a little over 2 percent of the local male workforce. Even in New York City, where nearly 55 percent of the male workforce was foreign-born, less than 34 percent of the chauffeurs were first-generation immigrants. Blacks constituted just over 5 percent of the chauffeurs in New York, while they accounted for only 2 percent of the city's male workforce (table 2).

Comparing these profiles to specific male work groups in given cities provides further insight into the identity of chauffeurs. Looking at Boston again, two census occupation categories recording similar numbers of male employees were street railroad conductors and "hostlers and stable hands."[25] The age and nativity/race profiles of streetcar conductors was very similar to that of chauffeurs, with conductors being just slightly older than chauffeurs and less likely to be black than either chauffeurs or the local male workforce. Hostlers and stable hands were as likely to be black as chauffeurs but much more likely to be first-generation immigrants. Male employees classified as domestic and personal servants in Boston were also much more likely to be foreign-born than were chauffeurs. Coachmen were counted among male domestic and personal servants, but their numbers were not differentiated in the occupation data for individual cities. Nationally, however, the 25,171 coachmen listed in 1910 were exclusively male, were older than chauffeurs, and were more likely to be foreign-born or black than were chauffeurs.[26] In general, we can safely say that nearly all chauffeurs were male, that they tended to be young, native-born whites with a minority of black practitioners, and that they were concentrated in the urban North. In these demographic respects they were more similar to streetcar conductors than to stable hands, servants, or coachmen.

TABLE 2.
Nativity and Race Profile of Chauffeurs Compared to All Employed Males in Major U.S. Cities, 1910
(Percent)

	Native White	Foreign-Born White	Black
Boston:			
Chauffeurs	61.0	32.9	5.8
Male Workers	49.4	47.6	2.4
Chicago:			
Chauffeurs	65.2	24.7	9.9
Male Workers	47.2	50.1	2.4
Los Angeles:			
Chauffeurs	75.9	14.1	8.5
Male Workers	54.5	24.7	1.9
New York:			
Chauffeurs	61.0	33.7	5.3
Male Workers	42.9	54.7	2.1
Philadelphia:			
Chauffeurs	64.7	17.9	17.3
Male Workers	61.0	33.0	5.8
San Francisco:			
Chauffeurs	78.0	20.4	1.6
Male Workers	53.1	40.1	0.5
St. Louis:			
Chauffeurs	63.5	13.0	23.3
Male Workers	67.2	25.4	7.3

SOURCE: Based on data drawn from U.S., *Thirteenth Census of the United States,* vol. 4: *1910,* 152–53, 181, 540, 545, 572, 575–78, 589.

NOTE: Percentages do not total 100 due to rounding and the exclusion of census category "Indian, Chinese, Japanese, and all other."

Judging from the public discussion generated by the chauffeur problem, chauffeurs developed their own ideas about their social status. Their knowledge of the new technology placed them in high demand, and this fact gave them, in their estimation, a status distinctly above the coachman of old and more akin to the railroad engineer.[27] A writer for *Automobile Magazine* agreed, finding that chauffeurs' intelligence compared favorably to that of coachmen: "looking at the chauffeurs . . . they are found to be bright, active, self-possessed, intelligent: they show, in short, the difference between a locomotive engineer and a horse jockey."[28] As a consequence, many chauffeurs refused to wear livery, some balked at requests to do menial jobs such as washing the car or announcing the car's arrival, and in Des Moines, Iowa, chauffeurs organized a strike against employers who required them to sleep in the same building with their machines.[29]

The question of status touched every aspect of motoring and touring. In hotels and inns along the New England motor touring corridor, motorists and other wealthy patrons often objected to chauffeurs dining in the same room with them—they were servants, after all. Yet when the chauffeur was shown to the servants' quarters, he became indignant about the treatment, complained to his employer, and refused to return to that establishment. Hotel owners were perplexed. They did not want to offend either the wealthy patrons or their chauffeurs. The wealthy motorists, in turn, were caught in a contradiction: they wanted a chauffeur who performed as a servant, but they did not want to offend and possibly lose a chauffeur who had the mechanical knowledge required to make such motor touring trips possible in the first place.[30]

Juvenile literature of the time also reflected this social ambiguity. C. N. and A. M. Williamson's *The Lightning Conductor: The Strange Adventures of a Motor-Car,* tells the story of Molly Randolph, daughter of a wealthy American businessman, and her tour of Europe with her Aunt Mary on a motorcar. (Before the widespread adoption of closed automobile bodies in the 1920s, motorists spoke and wrote of riding "on" an automobile rather than "in" an automobile.) Her first car was an unnamed double-chain drive German motorcar, and her first chauffeur was named "Rattray." When they stopped at a French inn, "the place was so thronged that Rattray had to sit at the same table with us," Molly wrote her father, "and though as a good democrat I oughtn't to have minded, I did squirm a little, for his manners—well, 'they're better not to dwell on.'" Rattray was hot-tempered and gloomy and eventually absconded with five hundred francs in repair money. The rest of the story revolves around the romance that develops between Molly and her second chauffeur, James Brown, actually one John Winston, "private gentleman and man at large, with a taste for travel," who offers to complete her tour on his Napier—"a snow-white car with scarlet cushions, all of the brass-work gleaming like a fireman's helmet"—and proves, by implication, that a true gentleman, who knows the social position of the chauffeur and is willing to play the role, can be a really good chauffeur. Eventually, Winston must reveal his true identity to Molly, as she could not consider a relationship with a mere chauffeur-mechanic.[31]

The conflict arising from the chauffeur's ambiguous and contested social status became acute in urban settings such as Boston, New York, and Philadelphia. The geography of city life added the final ingredient that allowed chauffeurs to alter substantially the power relationship between themselves and their employers. Unlike today's cars, which are fully enclosed, weatherproof, and climate controlled, early automobiles were mostly open-bodied and needed protection from inclement weather. In densely populated Eastern cities it was difficult for motorists to find accommodations for their automobiles which were in close proximity to their homes or businesses.[32] James Flink noted that before the introduction of the automobile, "the unsightliness and stench of the stable" led most urban horse owners to keep their horses at public livery stables "an inconvenient distance from their residences." The public automobile garage followed this precedent, partly because livery stable operators found it profitable to offer storage to motorists and partly because this was an urban building pattern that was already well established. High urban property values and high building densities left little room for private garages. Finally, fear of gasoline explosions and fires replaced the unsightliness and stench of the horse as reasons to keep vehicles stored away from urban residences.[33]

The earliest garages were invariably older buildings converted to the purpose, but eventually entrepreneurs built specially designed concrete or brick structures to house automobiles. Reflecting the high property values of the urban setting, most were two, three, and even four stories high, with elevators to move cars from floor to floor and turntables built into the floors to help maneuver cars in the tight quarters. Public garages offered a range of services to the motorist. Some simply offered storage, fuel, oil, water, and charging stations for electrics. Others offered to wash the customer's vehicle after every trip, sold parts and motoring accessories, rented cars and drivers for short trips, and performed regular maintenance and even major repairs. A number of the larger garages had recreation rooms set aside for the use of chauffeurs and shop areas for chauffeurs to perform their own work. The Decauville Garage, for example, provided a repair department "for the use of chauffeurs who wish to overhaul and repair their own machines." The Eureka Auto Station on West 124th Street in New York was laid out on a similar plan, with a "space devoted to repairs of cars by owners and chauffeurs" (see fig. 6).[34]

In 1905 practically all automobiles in large cities were stored in public garages, physically removed from their owners' presence, and because many wealthy motorists delegated responsibility for the care and maintenance of their vehicles to chauffeurs, the chauffeur usually became the principal negotiator for the vehicle's storage in the city. This situation gave the chauffeur the opportunity to dicker a bit with garage owners eager to fill their stalls. He could then play on the competition between these garages and garner anywhere from a 5 to 10 percent kickback on the monthly storage fees charged to his employer. Once the automobile had been placed, he might further bargain for a 5 to 15 percent commission on all of the gasoline, oil, parts, supplies, and services sold to his employer. Garage owners initially seemed willing to pay these fees to chauffeurs in order to attract business. Soon, however, they found themselves compelled to continue paying for fear that the chauffeur could easily sabotage the car with nicks, dings, and scratches, thus making it appear to his employer that the quality of service had degraded and the car was no longer being treated well at that garage. A chauffeur so inclined could then advise his employer to move the car to a garage he knew would be more cooperative and more forthcoming with the expected commissions.[35]

In receiving commissions, chauffeurs were not doing anything that coachmen had not done previously, often with the full knowledge and consent of their employers. Indeed, by demanding such commissions, chauffeurs attempted to retain and reinforce beneficial aspects of the coachman-based social structure at the same time they were challenging the servile status that employers were attempting to impose on them. Because many automobile owners were ignorant of and mystified by the workings of their machines, chauffeurs were able to exploit their authority over the vehicle to a much greater degree than had their predecessors. At a 1905 meeting of the New York City–based Automobile Club of America, Mr. Shattuck, a prominent member of the club, pointed out the historical precedent of paying commissions to coachmen and observed that the chauffeur "naturally drifted into a similar attitude toward the owner and dealer." Except, Shattuck noted, because many owners were not familiar with the workings of their automobiles, the chauffeur's "temptation to fraud has been very great."[36]

As corrupt as the commission system may have seemed to some, a more pernicious aspect of the chauffeur problem was the chauffeur's unauthorized use of the car. Exploiting his authority over the vehicle, a chauffeur might take his employer's car out, without permission, for joyrides about town. A chauffeur could also supplement his wages by as much as a hundred dollars per night by hiring himself and the car out as a sort of limousine service to downtown theater patrons, a practice called "hacking it." Some garage owners did not regard it as their business to regulate the comings and goings of chauffeurs to whom employers had delegated broad authority. For many of the same reasons that they felt compelled to pay commissions, they were inclined to look the other way when the chauffeur took the car out at night. Motorists often did not know that their car was out until it was spotted by a friend or until they received a call from the police saying it had been involved in a wreck.[37]

Resolving the Chauffeur Problem

Chauffeurs did not long maintain their advantage. Motoring and garage interests responded to the chauffeur problem by initiating legal, educational, and bureaucratic changes that severely restricted the chauffeur's power. Wealthy motorists particularly resented the joyriding aspect of the chauffeur problem; unable to prevent the practice, they worked to at least limit their own liability. In

1903 the Pennsylvania legislature passed the first laws regulating automobiles in that state. Six years later, at the height of the chauffeur problem, wealthy motorists shepherded a major revision through the legislature. They included a clause that legally distanced motorists from the consequences of joyriding chauffeurs. Section 23 of the motor vehicle act of 1909 mandated: "The registered number displayed on the motor-vehicle shall be prima facie evidence that the registered owner of said car was then operating the same: Provided, however, That [*sic*] if at any hearing or proceeding the owner shall testify . . . that he was not operating the car at the time of the alleged violation of this act, and shall submit himself to an examination as to who at the time was operating the car, and reveal the name of the person if known, then the prima facie evidence arising from the registered number shall be overcome and removed, and the burden of proof shifted."[38] Thus, Pennsylvania motorists could be absolved of responsibility for the damages done by their autos if they participated in the prosecution of their joyriding chauffeurs.

State and county court systems also supported wealthy motorists in their endeavors to protect themselves financially from joyriding chauffeurs. Between 1907 and 1913 a series of Pennsylvania court decisions established that anyone injured by a chauffeur-driven automobile in the state of Pennsylvania must show not only that the person driving was the owner's servant "but the further fact that he [the chauffeur] was engaged on the master's business, with the master's knowledge, and by the master's direction" in order to receive damages from the owner.[39] These decisions ensured that motorists could not be held liable for their joyriding chauffeurs. New York courts went further. In an attempt to sever the relationship between chauffeurs and garage managers believed to be at the root of the urban joyriding problem, New York courts began finding garage owners liable for the damage done to cars taken out without the motorist's permission. Similarly, in 1910 a Kansas jury found a garage owner, a garage employee, and the joyriding chauffeur all liable for the damage done to a judge's car taken from a Topeka garage.[40]

Parallel to these measures, New York and other states initiated statewide testing and licensing of all chauffeurs. The Pennsylvania law of 1909 repealed the previous requirement that all motor vehicle operators obtain a state license. Rather, it required only that "every person desiring to operate a motor-vehicle as a chauffeur, or as a paid operator, shall first obtain a driver's license."[41] Chauffeurs, moreover, had to be over eighteen years of age and were required to wear a two and a half–inch badge with their number and the words *Pennsylvania Licensed Driver* on the front of their outer garment whenever they drove.[42]

Wealthy motorists also encouraged the YMCA to get involved in trying to remedy the chauffeur problem. In an attempt to restore employers to an advantaged position, the Boston and New York YMCAs offered automobile courses for owners and for chauffeurs. The West Side Branch of the New York YMCA opened its automobile school in 1904 with the aid of a one thousand–dollar gift from New York's socially elite Automobile Club of America. The school's first brochure stated, "The aim of these courses will be two-fold: first, to train and supply competent operators and chauffeurs; and, second, to give owners and prospective owners a sufficient knowledge of the theory and practice of automobiles and automobiling to enable them to meet the emergencies that constantly rise in connection, not only with machines, but with chauffeurs."[43] The YMCA's program sought to decrease motorists' dependence on chauffeurs while at the same time alleviating the shortage of young men with mechanical knowledge of the automobile. Wealthy motorists and soon-to-be motorists attended YMCA lectures to familiarize themselves with the principles of motorcar construction and operation. Prospective chauffeurs learned how to drive, maintain, and repair touring class automobiles and, ideally, how to behave "professionally."[44] Chauffeurs of "approved character" with "satisfactory references" who passed the practical exams at the end of the course were granted a chauffeur's certificate and were eligible to use the YMCA's Employment Bureau to help secure a suitable job.[45] The West Side YMCA's automobile courses were hugely popular. They maintained consistently higher enrollments and generated more income than any of the YMCA's other trade and business courses. During the first ten years of the auto school's operation, the West Side YMCA trained over 10,500 New Yorkers through its auto school, almost half of them as chauffeurs.[46]

Education thus helped resolve the chauffeur problem for some wealthy motorists, who were able to retrain their coachmen as mechanics successfully, hire reliable school-trained chauffeurs, or become themselves technically competent supervisors. One motorist who paid for a boy's training at the New York School of Automobile Engineering wrote to a trade publication complaining of the typical problems with other chauffeurs he had employed: surliness, dishonesty, "incompetence." Then he had the bright idea "to take a young man whose parents had been working for me for years and break him into automobile work, thinking that it might result in my getting for a chauffeur a man who knew his position and give [*sic*] me the credit for being the boss." He wrote to express his "perfect satisfaction" with his school-trained chauffeur: "He realizes that I understand the mechanism as well as he does—I took a course in the school myself—and keeps the repair and supply bills to a point that seems ridiculous after what I have been used to . . . the relief of mind is inexpressible." The wealthy motorist maintained a paternalistic control over the young man and his parents which ensured the boy's deferential behavior. The automobile school enabled him to purchase the technical training needed to turn the boy into a chauffeur-mechanic and himself into a knowledgeable supervisor, replicating the familiar master-servant social structure of elite horse transportation.[47]

Garage owners focused their attention on the kickback and commission system that some chauffeurs were so effectively exploiting by mid-decade. Motorists, by and large, seemed none too concerned about their chauffeurs receiving a little extra income through such commissions. They wanted their chauffeurs to be content. Like garage managers, they too knew what troubles a disgruntled chauffeur could cause for their machines. Motorists also seemed to have understood that if their chauffeurs were denied their extra income, there would be little chance the savings would be reflected in lower prices or storage rates. Consequently, the most vocal complaints about the commission system came not from motorists but from the garage interests out of whose profits the commissions were paid.[48]

Garage managers, therefore, sought in their own way to restrict the chauffeurs' power over their relationship. Once dependent upon chauffeurs' for business, garage managers began to resent the drain on profits which the commission system represented. They attempted to gain the upper hand with chauffeurs by forming alliances with other garages. Members would agree not to succumb to the temptation to lure the chauffeur's business with commissions on sales and storage. So organized, they could not be played off against one another by chauffeurs in search of higher commissions and greater privileges, as had formerly been the case.[49] Competition and distrust among garage owners, however, forestalled an effective alliance among New York garages until the end of the decade.[50]

Some garage managers sought to circumscribe chauffeurs' activities through bureaucratic means. The pages of *Horseless Age* reflect a flurry of activity by garage managers between 1908 and 1909. They developed duplicate and triplicate forms and accounting procedures designed to establish direct communication with the owner about charges for fuel, oil, supplies, and repairs.[51] By involving the owner directly in decisions about the care of the vehicle and by conducting all financial business through the mail, garage managers hoped to circumvent the chauffeur's authority over the vehicle. One writer for *Horseless Age* recommended establishing a citywide blacklist of New York chauffeurs who demanded commissions.[52] Historical evidence of such a blacklist has not surfaced; the Philadelphia Automobile Club, which ran its own garage in downtown Philadelphia, did, however, establish what they called a "Chauffeur Bureau" for the use of members. Former employers of chauffeurs seeking work were asked such questions as "What wages did you pay him?" "Is he a good mechanic?" "Is he honest?" "Is he sober?" and "Is he willing?" Presumably, if a chauffeur received commissions from garages or suppliers, he might fail the honesty question, and if he did not behave deferentially, he would fail the willingness question.[53]

Responding to pressure from motorists to curb the practice of joyriding, reform-minded garage managers further circumscribed the freedom of chauffeurs. One garage in New York stationed a young man on a balcony over the main floor and paid him to track the movements of each car and chauffeur by use of a time clock. That information was then mailed directly to the automobile owner at his residence or place of business.[54] The National District Telegraph Company offered garages an early form of electronic surveillance. Each stall would be equipped with a receptacle, and cars would be plugged and unplugged from it upon arrival and departure, making a telltale mark on a twenty-four-hour register kept under lock and key in the garage's office.[55] By 1909 the editor of *Horseless Age* surmised that most of the garages in the eastern states had "a very complete checking system, covering the leaving and returning of all cars, all work done on cars, supplies furnished, etc."[56] Due to the concerted efforts of garage owners and wealthy motorists, by the end of the decade chauffeuring had become a very closely monitored, scrutinized, and regulated occupation.

In addition to these legal, educational, and bureaucratic reform measures, a steady undercurrent of technological change undercut the chauffeurs' mechanical authority. In order to appeal to a broader market, many American automakers began to focus on producing moderate- and low-priced cars that would meet the needs of businessmen, tradesmen, and farmers. These new motorists could not afford, and did not desire, the services of a chauffeur-mechanic and pushed the American automobile industry to produce more reliable cars for owners in all social classes. As reliability increased, the technical demands that initially had favored the use of chauffeurs decreased. The development and proliferation of the demountable rim, for example, made blowouts less troublesome. One motorist, writing to his son after attending the 1910 New York Auto Show, exclaimed that "the Goodyear detachable-demountable tire was a wonder. You merely pushed the whole thing on the rim, gave it a kick, and it was all right. Some springs held it in position. Nothing could be simpler and easier."[57] Drive mechanisms became more reliable as trouble-prone chain drives gave way to drive shafts, better roads did less damage to springs and frames, and myriad other technical developments such as improved ignition systems, better engine block castings, and more effective lubrication systems diminished the need for drivers to double as mechanics. After 1910 automobile trips of one hundred miles and more per day, without incident, became commonplace. No longer did manufacturers have to advertise as the Cadillac Company did in 1903: "When you buy a Cadillac, you buy a round trip." In this changing technical climate the wealthy motorist no longer needed to have his mechanic on board at all times.[58]

For their part chauffeurs did not begin to organize effectively until late in the first decade of the twentieth century. Even then, if the activities of National Chauffeurs' Association are any indication, they did not attempt to protect their authority over the vehicle or the unique technological knowledge on which their earlier authority was based. Instead, they worked toward "the uplift and betterment of chauffeurs" through the sponsorship of yet stricter licensing laws and the advocacy of state-sponsored employment exchanges and apprenticeship programs. Displaying one of the baser tendencies of workers who feel themselves losing power, they also attempted to purge their profession of blacks and foreigners. The National Chauffeurs' Association denied membership to "Negro drivers," and some chauffeurs were suspected of sabotaging cars driven by black chauffeurs in an attempt to discourage their hire by motorists.[59] Another organization, the Chauffeurs' Professional Club of America, apparently introduced legislation in New York which, in addition to tightening licensing requirements, also prohibited aliens from obtaining a chauffeur's license.[60] These efforts appear to have been only partially effective in determining who could and could not become a chauffeur.[61] Despite their late organizing attempts, chauffeurs never regained their former power—either over the machine or vis-à-vis their employer or the garage manager. They were never again able to "lord it over their employers," as the *New York Times* had noted with scandalized alarm in 1906.

The number of chauffeurs continued to increase, and by 1920 over 285,000 were employed nationwide—a more than sixfold increase within a decade. The chauffeurs of the second and third decades of the twentieth century were professional drivers, not mechanics, however, and their social position as servants was relatively stable and uncontested. In 1911 a contributor to *Horseless Age* provided an early description of the daily routine of this new type of chauffeur. He started his day at eight o'clock in the morning by getting the car ready for the day's use. At nine he reported for duty to drive his employer to work. Returning to the house at mid-morning, he would take the ladies of the house "on a tour of the shops" and to a fashionable restaurant for lunch. Around two-thirty or three in the afternoon he would take the children for a drive. After dinner he would drive his employer and his wife to the theater or opera, returning the car to the garage between one and two in the morning. Significantly, the author made no mention of the chauffeur performing any mechanical duties other than "preparing or overseeing the preparation of the machine for the day's use."[62]

There was no pronouncement in the *New York Times* or in the automotive trade press that the chauffeur problem was over, but discussion of it virtually disappeared after 1914, fulfilling the prediction of one *Motor World* editor that "in the fullness of time, conditions will have changed so completely and yet so imperceptibly that many will wonder what there was to be alarmed about."[63] The discursive quietude that followed heralded an emerging, alternative equilibrium. As roadside repairs became less frequent and automobile ownership became more commonplace, motorists of all classes chose to have their machines maintained and repaired at the dealer's shop, the independent garage, and eventually the corner gas station, and chauffeurs became the professional paid drivers we think of today.

Despite the resolution of the chauffeur problem, motorists continued to have ambiguous and conflicting attitudes toward mechanics. A 1917 Hudson *Service Inspection Manual* reminded Hudson owners "that almost every man who owns an automobile has accumulated a little more than the average amount of wealth. . . . Therefore, he is very apt to possess ability beyond the average found in the working classes, from which ninety percent of our mechanics are drawn. . . . To expect all mechanics to display intelligence, judgment and foresight equal to that of the average Hudson owner is unreasonable."[64] In language reminiscent of that used to describe coachmen and chauffeurs, the Hudson Motor Car Company appealed to motorists to be generous and patient when dealing with garage mechanics. Yet in the absence of day-to-day social contact with mechanics, as had been the case with chauffeurs, motorists did not feel as much pressure to define clearly their social relation with mechanics. A brusque or even surly mechanic need only be endured for the limited time of the repair or service needed—not for the duration of a New England motor tour. Resolving the conflict between class, knowledge, and power in the repair shop was less pressing under the new structure of personal transportation. Likewise, the technological knowledge that gave chauffeurs their short-lived power did not disappear along with the chauffeur problem. The locus and control of specialized repair knowledge merely shifted to the independent and dealer repair shops, where mechanics continued to wield their authority and power in opposition to perceived injustices with varying degrees of success.

CHAPTER 2

Ad Hoc Mechanics

Accounts of early motorists often mention the important role ad hoc mechanics had in keeping these new machines in repair. Floyd Clymer recalled that in his boyhood town of Berthoud, Colorado, about 1906–7, "Our car engines were sometimes repaired by the local plumber, Andy Bergun—and, when the springs broke on the rough dirt roads (which was quite often) blacksmiths Bimson or Preston would do a nice job of welding for about a dollar per spring."[1] Bellamy Partridge wrote of driving his first automobile, a two-cylinder Rambler, on an ambitious 112-mile trip in about 1905–6. The Rambler was a lightweight, tube-frame type buggy rather than a touring car, and Partridge did not employ a chauffeur. Besides stopping four times on his trip for tire troubles, he stopped once at a blacksmith shop to have a broken fender support welded for fifteen cents. "A little later," he recalled, "I limped into Batavia [New York] with the engine missing badly, or 'skipping,' as it was then called. The man at a bicycle shop suggested a new set of dry cells [batteries] which, he said, had solved the difficulties of another car only a few days before. He installed a new set, six of them, and we drove on to Buffalo without further trouble, except that the driving chain came off once."[2] These accounts typify how early motorists turned to workers they thought could help them, apparently based on their assessment of the individual's general mechanical ability or on their physical proximity at the time of need. Few motorists today would entrust their automobile to a plumber, but in 1906 a good plumber who was curious about mechanical things might well have been just as qualified to repair an engine as any other person in a small town.

During the period when wealthy motorists employed private chauffeur-mechanics, increasing numbers of other workers and tradesmen began servicing and repairing automobiles on an ad hoc basis. They came from diverse backgrounds and brought a range of technological experiences and social contexts to their work on automobiles. Machinists, blacksmiths, bicycle mechanics, electricians, patent model makers, carriage makers, plumbers, and assorted others began learning about, tinkering with, and repairing automobiles as the technology came into wider use. These miscellaneous workers were ad hoc auto mechanics in the sense that they took up auto repair work as a sideline to their other work. They were ad hoc also in the sense that they were the ones who were available to motorists who did not want to employ a chauffeur-mechanic.

Ad hoc auto mechanics did not generally set out to become auto mechanics. Rather, motorists turned to them when they needed help. Initially engaged in other trades, ad hoc mechanics took on automobile work with varying degrees of enthusiasm. Certain workers, however, had technological knowledge and resources that were readily transferable to early automobile repair work. Yet because automobiles incorporated different kinds of technology into a single complex mechanical system, no single trade or occupational group enjoyed a monopoly on early auto repair work. In fact, a closer examination of the turn-of-the-century blacksmith, the classic example of the ad hoc mechanic, reveals that technical competence was only one factor in making the transition to automobile mechanic. Long-established economic and social relationships may have prevented many ad hoc mechanics from becoming full-time automobile mechanics, while technical curiosity and youthful enthusiasm drew others into the trade.

Motoring without a Chauffeur

Partridge's experience in Batavia highlights one of the lesser-known groups of ad hoc mechanics: bicycle mechanics. The transfer of bicycle manufacturing knowledge to early automobile manufacturing has been well documented by historians of technology.[3] Less well-known is the parallel translation of technological knowledge on the repair level. Many of the early non-chauffeured automobiles were constructed, as Partridge's Rambler, of bicycle-like tube frames and wire-spoked pneumatic tires. The knowledge, experience, and even tools of the bicycle mechanic easily transferred to work on these types of automobiles. This translation took physical form in the Young and Company automobile garage in Riverside, California. Cornelius Young ran two businesses at the corner of Orange Street and Eighth Street: an automobile garage at 768 Orange Street and a bicycle shop at 587 Eighth Street. The two properties joined in a common shop at the rear, where Young could work on either automobiles or bicycles.[4]

In another example Sydney Bowman established a bicycle sales and service shop in 1891 at the corner of Eighth Avenue and Fifty-sixth Street in New York City. The Sydney B. Bowman Cycle Company's 1901 brochure announced that they were expanding their line to include automobiles, motorcycles, phonographs, launches [motor boats], and skates. "Our workshop," bragged the brochure, "is complete in all its details, and we can undertake and successfully perform any job, from repairing the comparatively heavy mechanism of an automobile or building a bicycle, to adjusting the delicate and intricate parts of a phonograph." Bowman evidently had more success and sensed greater demand for his knowledge and experience in automobile work than in boating or phonography. A brochure from the next year announced simply the Sydney B. Bowman Automobile Company, a "sale, storage and repair station," at 52 West Forty-third Street.[5]

Far from being isolated examples, the transition experiences of Young and Bowman were reflected in the pages of the bicycle trade press of the time. The journal *Bicycling World* appended "and Motorcycle Review" to its title in late 1900 and urged readers to prepare for "the coming of the motorcycle. . . . [M]en who will survive the coming struggle . . . are studying motorcycles, dissecting them, repairing them at nominal figures . . . anything to acquire experience" (see fig. 7).[6] An article the following year illustrated and described the mechanism of a one-cylinder, gasoline-powered, three-wheeled motorcycle.[7] Little technical difference distinguished early motorcycles from light gasoline buggies such as the Rambler and the Oldsmobile, so this information likely proved helpful to other bicycle mechanics similar to Young and Bowman, particularly since the bicycle craze of the late nineteenth century was quickly fading, and many may have felt pressure to leave the bicycle trade. A popular recreational monthly of the time, *Outing Magazine,* covered automobile news as a department of sports. This association may have also led a number of motorists to the door of the bicycle shop for accessories, directions, maps, and of course repairs. When stranded motorists such as Partridge sought help at bicycle shops, they likely found eager and able mechanics.

Yet ad hoc mechanics could come from any number of backgrounds, and during the first decade of motoring there were no established criteria for choosing a mechanic. When the early motorists in Washington, D.C., experienced troubles with their automobiles, some turned to Frederick Carl, a precision machinist. In 1890 Carl established himself in the nation's capital as a model and precision instrument maker. As automobiles became popular, Carl found himself making patent models for inventors of carburetors, electric cars, and other parts. When motorists familiar with Carl's model-making expertise had trouble with their cars, they sought his help on their full-sized models as well, and by the mid-1930s his automobile service business had grown under the management of his four sons to be the largest in the District of Columbia.[8]

As these examples indicate, ad hoc mechanics from the metal trades enjoyed a distinct advantage when working on early automobiles. Because much of the repair work on early automobiles involved fabricating, adjusting, or repairing metal parts, machinists, blacksmiths, bicycle mechanics, and even plumbers often found their services in demand by motorists stranded on the road. These workers already possessed some of the technological knowledge required to be good mechanics: what one blacksmithing journal called "the knowing how tight to turn a nut, the strength of material, and how much play should be allowed in a bearing."[9] General machinists and smiths could employ or transfer their knowledge of materials to fashion a new or repaired part that often surpassed the performance of the original. Yet these were not the only or even the most important skills needed to become a good automobile mechanic.[10] Oscar Friedrich, a New York blacksmith, cautioned fellow smiths, "The automobile trade compared with the blacksmith trade is like a watchmaker compared to a toolmaker," in that automobile work required experience with finer mechanisms and different systems than the average smith possessed.[11]

Most ad hoc mechanics probably learned about automobile technology through a combination of tinkering and reading. The challenge to those who learned by tinkering came in identifying engine problems related to the carburetor and ignition systems of the rapidly proliferating gasoline-powered automobiles. Some from the metal trades may have learned about such systems if they employed stationary gasoline engines in their shop or if they worked closely with a customer who used them. W. A. Rickert, a Kansas blacksmith, gained some practical knowledge of air-fuel ratios when using his two-horsepower gasoline engine during cold winter weather. "To start and run it according to directions was impossible," he wrote. "But I found that by putting a piece of corn cob in the mouth of the air pipe to shut off nearly all the air at the start, and giving the engine more air as it got warmed up, that it started as easily as in warm weather." Rickert thus invented a manual, corncob choke for his carburetor.[12] Such prior experience certainly would have provided the tinkerer with a leg up. Yet diagnosing problems with engine performance required knowing something of the principles of internal combustion engines, and learning such principles exclusively through empirical exploration would have been exceptional.

For the unexceptional and the inexperienced, printed material supplemented their prior knowledge and exploratory methods. Ad hoc mechanics could read the growing number of trade journals, books, and manufacturers' manuals available after the turn of the century. *Horseless Age*, the automobile industry's first trade journal, debuted in 1895 and published a range of articles aimed at automakers, automobile owners, and those who worked on autos. It is doubtful, however, if many early copies of this journal made it into the hands of ad hoc mechanics. They more likely found automotive information in their own primary trade journals. *American Blacksmith, American Machinist, Bicycle World, Blacksmith and Wheelwright,* and the *Hub* all published technical articles about automobile repairs and general articles on the principles of internal combustion engines.[13] In addition, publishers placed ads in these journals promoting a variety of newly available automobile books, such as those by Leonard Elliott Brookes, Andrew Lee Dyke, Victor Pagé, and Charles P. Root. Ad hoc mechanics could consult these books for general and detailed information.[14] Finally, some early automakers included considerable mechanical detail in their owners' manuals. It was not uncommon for the stranded motorists to let the machinist or blacksmith read the manual before undertaking any major repair work.

Workers from a variety of backgrounds, then, could transfer and supplement their technological knowledge in order to step in and assist early motorists who did not employ their own chauffeur-mechanics. Moving from ad hoc status to full-time automobile mechanic was not, however, simply a matter of possessing the requisite technological knowledge, experience, or tools. A closer look at the stereotypical ad hoc mechanic, the blacksmith, illustrates how various other factors affected the transition from ad hoc to full-time status.

The Village Blacksmith

John Vander Voort worked as a blacksmith in rural Hunderton County, New Jersey. On 4 June 1908 one of his regular customers, J. N. Pidcock, paid him seventy-five cents to repair the "friction band," or clutch, on his automobile. This marked only the second time that Vander Voort had repaired an automobile in his shop. The first was in the previous fall, when he repaired William Taytor's drive chain for the sum of one dollar. Pidcock returned within two weeks to have more work done on his clutch. Over that summer Vander Voort did more automobile work and became, in retrospect, an ad hoc auto mechanic. By the end of the motoring season Pidcock had brought his automobile to Vander Voort six more times for various minor repairs. Thus begins what one would assume to be a classic account of a turn-of-the-century village blacksmith making the logical career transition to automobile mechanic. Yet Vander Voort never became a full-time automobile mechanic. His daybook entries indicate that he continued to make repairs to Pidcock's automobile and to a handful of others through the 1910s. Beginning in 1920, however, his only regular automobile customer was the Union Garage, which subcontracted heavy metalwork jobs to him such as straightening axles and welding engine cylinders. Never did Vander Voort record in his daybooks that he adjusted a carburetor or rebuilt an engine, and his daybooks end in 1923 with no indication that his business grew beyond blacksmithing.[15]

Vander Voort's experience reveals that not all ad hoc mechanics made the transition to full-time mechanics and calls into question the widely held assumption that a large number of early automobile mechanics came from the ranks of blacksmiths. This notion has persisted through the years partly because of the numerous anecdotal accounts of early motorists stopping at a country blacksmith shop for this or that repair and partly because some blacksmiths did indeed become full-time auto mechanics. Almost every local or county historian can point to an automobile dealer or repair shop in his or her community that began as a blacksmith's shop. Yet the notion that many or most blacksmiths became auto mechanics seems to have been generally accepted and never critically examined.[16]

One historian has questioned whether blacksmiths had the requisite knowledge to repair this new machine. "While blacksmiths could straighten a bent axle or forge a new leaf spring, there is little reason to think they could adjust the vibrators of an early ignition system, rebuild the main bearings, or even repair a leaking radiator as well as could other kinds of tradesmen." Bicycle shop owners, electricians, machinists, livery stable operators, and even hardware dealers "were just as likely to work on cars and become auto mechanics" as were blacksmiths.[17] Ad hoc mechanics did indeed have diverse origins, though another scholar studying the automobile repair industry has countered that "blacksmiths entered the automobile trade in large numbers not merely for technical reasons, but also for social reasons. Because of their centrality in village and small town life people were simply used to getting various items 'fixed' by the local blacksmith in a way that they were not used to visiting a machine shop or an electrician. Furthermore, blacksmiths may have had more reason to consider changing trades than machinists and electricians, both of which were trades with higher wages than those of automobile mechanics."[18] Neither scholar delved deeply into this question, as their main concerns lay elsewhere. Yet such questions are important for understanding the origins of the auto mechanic's trade.

Vander Voort's daybooks support the notion that the blacksmith's preexisting social role in small communities encouraged customers to bring their cars to his shop for repair. Most of his automobile customers were regulars in his shop before they brought in their cars. In a society reliant upon horse-drawn vehicles, the blacksmith, wheelwright, and carriage makers' shops were already the locus of most of the repair work associated with personal transportation as well as farm and household goods. Blacksmiths welded broken metal, forged new parts, shoed horses, repaired tools, and sharpened plows. Wheelwrights and small-time carriage makers performed a variety of wood-related repair tasks such as replacing cracked boards in carriage bodies and broken spokes in wheels. In small rural villages these metal and woodworking abilities were often embodied in a single person, or when they specialized, blacksmiths, wheelwrights, and carriage makers often worked in close harmony with one another, perhaps even in partnership under the same roof. Whatever the case, customers from the local community expected that they could get knowledgeable and reliable work from these tradesmen.[19]

Yet that same community identity could hamper the blacksmith's ability or willingness to take up automobile work. In some rural settings the first motorists to request the blacksmith's help were not his regular customers, nor even members of the community, but wealthy, urban motorists out for a tour of the country. These blacksmiths found that motorists and farmers were initially two different and sometimes antagonistic constituencies. Long-standing social and economic ties with the farmer discouraged smiths in this situation from seeking regular automobile work. Eventually, farmers began buying their own automobiles, especially the Ford Model T, but, ironically, this did not increase the likelihood that the local blacksmith would become a full-time auto mechanic. Timing, geography, competition, technical knowledge, and the mechanical aptitude of farmers could combine to dissuade blacksmiths from making the transition to full-time auto mechanic even after millions of Model Ts washed over the countryside. To understand these developments better we must look more closely at the social and economic role of the blacksmith in the rural community.

Blacksmiths enjoy an almost legendary status in American history. In 1840 a young Henry Wadsworth Longfellow penned his famous poem "The Village Blacksmith," and its popularity during the late nineteenth and early twentieth centuries reinforced in the minds of subsequent generations of schoolchildren a belief in the yeoman dignity of the steady, earnest blacksmith:

> Under a spreading chestnut-tree the village smithy stands;
> The smith, a mighty man is he, with large and sinewy hands;
> And the muscles of his brawny arms are strong as iron bands.
>
> His hair is crisp, and black, and long, his face is like the tan;
> His brow is wet with honest sweat, he earns whate'er he can,
> And looks the whole world in the face, for he owes not any man.
>
> Week in, week out, from morn till night, you can hear his bellows blow;
> You can hear him swing his heavy sledge, with measured beat and slow,
> Like a sexton ringing the village bell, when the evening sun is low.

And children coming home from school look in at the open door;
They love to see the flaming forge, and hear the bellows roar,
And catch the burning sparks that fly like chaff from a threshing-floor.

He goes on Sunday to the church, and sits among his boys;
He hears the parson pray and preach, he hears his daughter's voice,
Singing in the village choir, and it makes his heart rejoice.

It sounds to him like her mother's voice, singing in Paradise!
He needs must think of her once more, how in the grave she lies;
And with his hard, rough hand he wipes a tear out of his eyes.

Toiling,—rejoicing,—sorrowing, onward through life he goes;
Each morning sees some task begin, each evening sees it close;
Something attempted, something done, has earned a night's repose.

Thanks, thanks to thee, my worthy friend, for the lesson thou hast taught!
Thus at the flaming forge of life our fortunes must be wrought;
Thus on its sounding anvil shaped each burning deed and thought.

Despite such romantic notions about the blacksmith's trade, by the eve of the twentieth century blacksmiths in the United States had seen important aspects of traditional metalworking leave their shops to become industrialized specialties. For a time in the late nineteenth century the line between blacksmith and machinist was unclear.[20] Yet by the turn of the century foundry men and industrial steelworkers performed the magic of turning hot, molten iron into objects of heft and importance, and machinists worked cold iron and steel into the intricate mechanisms of modernity. Blacksmiths consequently found their business narrowing to the repair of broken metal goods. Industrialization also affected the nature of work that remained in the blacksmith's domain. The increasing availability of drop-forged carriage hardware encouraged some blacksmiths simply to replace broken iron parts, which they had formerly welded, repaired, or fabricated from bar stock. Machine-made horseshoes, widely available after the mid-nineteenth century, meant that rural smiths could purchase standard shoes in quantity and then simply modify or alter them for their customers. The proto–mass production of wooden carriage parts such as wheels and spokes also encouraged replacement over fabrication or repair.

Nevertheless, the blacksmith's key role in the rural community remained largely unaffected. Farmers yearly increased their dependence on the new machinery of agriculture: steel plows, cultivators, combination harvesters, and even the occasional steam traction engine or stationary gas engine.[21] Under the increasingly mechanized and heavily mortgaged farming conditions of the late nineteenth century, broken equipment could be devastating. One Dakota farm wife wrote in 1889 that when a major part of a large steam threshing unit could not be repaired promptly, "then men must be laid off, the meat and pies in the pantry [for the hired men] spoil, rains come to destroy the grain in the fields and the yields at threshing time are reduced. This means more debts, hardships and discouragement." She underscored the farmer's dependence on machinery with her anxious recognition that "threshing day means more than noise, bustle, and dirt; it is the day when the farmer's balance sheet is made out. It is the day when the question of profit or loss for the year is determined."[22] This dependence on expensive agricultural machinery was especially acute in the grain-growing regions of the Midwest and Pacific states, but farmers in every region of the United States used more machinery on the eve of the twentieth century than a generation earlier (see fig. 8).[23]

The increased mechanization of farming helped offset the effects of industrialization for many blacksmiths. Their jack-of-all-trades repair knowledge could be crucial to the farmer, and unlike stereotypical factory workers, turn-of-the-century blacksmiths enjoyed a high degree of daily and seasonal variation in their work routines. Such variation stands out as the most salient feature of Vander Voort's daybooks from 1906 to 1917. In addition to shoeing horses and setting tires—by far the most common of a blacksmith's routine work—Vander Voort's work on agricultural machinery included sharpening harrow teeth, making braces for a corn plow, filing mower guards, welding a bolt for a press, sharpening corn plowshares, making new share bolts, repairing a harrow frame, repairing a fodder cutter, repairing a corn plow iron, and repairing the wheel on a traction engine. His repairs to horse-drawn vehicles included fabricating a new clip on a light axle, replacing body irons, welding a whiffletree clip, installing a new spoke in a wheel, and changing a buggy spring. He also repaired a rocking chair, made iron bars for the windows of a stable, repaired an emery grinder, fabricated a new *S*-wrench, and sold "colic cure," "gall cure," and "linement" [*sic*], presumably for ailing horses rather than ailing customers.[24]

Vander Voort was not alone in performing a variety of tasks in his shop. Tool lists, price lists, and floor plans submitted by readers to *Blacksmith and Wheelwright* indicate the varied nature of the blacksmith's trade at the turn of the twentieth century. Published by M. T. Richardson from January 1880 through September 1932, *Blacksmith and Wheelwright* targeted the small-shop reader and regularly published letters, questions, and tips from practicing blacksmiths.[25]

In February 1898 a Louisiana reader asked what tools would be necessary to open a small blacksmith shop. Two readers responded with lists of what they considered were the essentials for such an undertaking. Stanton Hoover, proprietor of "a small shop in the country" (Croton, Ohio) which took in "all kinds of farm, plow and wagon work, horseshoeing, etc.," listed thirty-two items he thought were "absolutely necessary." L. P. Schell, of Caton Farm, Illinois, listed fifty-one tools "necessary for a country blacksmith and wheelwright shop." The two lists overlap considerably and highlight the varied nature of the blacksmith's work: tuyere, bellows, anvil, hammers, and tongs for metalwork; coachmaker's vise, hand planes, spoke shaves, and hand saws for wagon and woodwork; shoeing box, pincers, hoof shears, nippers, and rasps for horseshoeing.[26] Price lists, submitted in response to editorial pleas to raise prices in the trade, provide another measure of what tasks blacksmiths considered routine enough to warrant a fixed price: sharpening various plowshares, lister shares, and cultivators; setting wagon and buggy tires; horseshoeing; replacing spokes and felloes; welding and replacing wagon springs; and making new wagon tongues, wagon beds, and buggy shafts.[27]

Finally, two shop floor plans submitted to *Blacksmith and Wheelwright* in March 1897 provide additional insight into the rural blacksmith's work and role in the community. The magazine published these plans as examples of "modern" shops because they illustrated the use of steam engines to drive much of the shop equipment. The machinery and layout of each shop nonetheless conforms to the range of tasks noted earlier. Each plan included an area set aside for processing agricultural products from the local community. Frank Shaw's plan identified a "feed mill," and Swarer and Robinson's plan depicted a sausage chopper and a sausage stuffer, which they used in the fall and winter to "chop the farmer's sausages." Thus, even in such modern shops blacksmiths strove to meet the needs of their rural clientele and neighbors.[28]

Vander Voort was similarly integrated into the agricultural context of his community. Most of his clients were regular, repeat customers, and with some he maintained a barter-like system of payment. One customer paid on a regular basis with "milk tickets." Others paid with oats, straw, or hay. In the entries for late June 1908 one finds payment "by tomatoes .20[¢]," and "by 20 quarts blackberries [$]1.60." Such entries reveal a close, familiar, and trusting relationship between Vander Voort and his customers. Together with the repairs he made to the farmers' equipment, they suggest that he was professionally integrated and personally vested in the agricultural life of the community. His custom ebbed and flowed with the agricultural season, just as the commodities he accepted as pay fluctuated with the seasons. Vander Voort appears not to have been rich nor to have been a significant landowner, but he filled an important role in the community.

This agricultural orientation of many village blacksmiths accounts not only for the wide variety of tasks they undertook but also goes a long way toward explaining why many of them did not fully make the transition from horse-drawn to horseless vehicle repair, even though it appears in hindsight to have been logical. For many blacksmiths their position in the community was relatively secure; they did not need to take up auto repair at an early date to remain solvent. Automobiles did not abruptly replace horses from the blacksmith's point of view. In 1908 Americans still used an estimated twenty million farm horses and untold thousands of plows, cultivators, and the like.[29] In short, blacksmiths had plenty of work to keep themselves busy without going after automobile traffic. Some even explicitly rejected automobile work, despite the urgings of industry sages, because it would distract them from their regular, agriculture-based work.

The Automobile in the Blacksmith's Shop

Any blacksmith or carriage maker who subscribed to or read an industry journal during the first two decades of the twentieth century would likely have felt besieged by advisors pushing automobile work as the key to his future. Despite initial reassurances by many carriage industry leaders that the horse would not be dethroned, the trend toward motor vehicles soon became apparent to all but the most sheltered tradesmen. Richardson, writing from his New York City–based offices, began pushing automobile repair work years before many of his *Blacksmith and Wheelwright* readers had even seen an automobile. In January 1900 he predicted—or prescribed—that as soon as automobiles came into general use, "the blacksmith will study their construction and prepare himself to make repairs."[30] In late 1901 he noted that a "great many" blacksmiths were profiting from bicycle repairs, then he rhetorically asked his readers, "why not automobiles?"[31] Thereafter, his journal relentlessly urged blacksmiths to take up automobile repair work. Between 1901 and the United States's entry into World War I, Richardson's journal published more than sixty-five editorials, testimonials, and advice columns about the advantages of taking up automobile work. In addition, regular readers could not have missed the increasing number of advertisements oriented toward automobile repair and supplies.[32] Just as the *Hub,* a journal aimed at medium to large carriage makers, had asserted that automobile carriage work rightfully belonged to the carriage maker, so did *Blacksmith and Wheelwright* assert that auto repair work "properly belongs to the blacksmith."[33]

It also cautioned blacksmiths who did not pursue automobile work: "If you are not [doing so], we are afraid you are making a mistake and you are likely to lose a large increase in your future business."[34]

Blacksmith and Wheelwright preached a sermon of salvation through auto repair in various guises. The earliest appeals recognized the class and regional distinctions between rural blacksmiths and early motorists and in not-so-subtle language urged blacksmiths to take advantage of the fleeting power that breakdowns presented. Because the first automobiles that readers of the journal would have encountered likely would have been owned by wealthy urbanites out for a tour of the country, "the nearest blacksmith [was] pretty sure to get a job" when the machine broke down. This would be doubly fortuitous, for "automobile owners generally are men of means and are willing and able to pay a good price for repair work."[35] In other words, blacksmiths who had prepared themselves by learning all they could about automobiles and who had obtained a few special tools could make some quick cash off the rich city folk. "Such jobs may not be numerous at present," conceded *Blacksmith and Wheelwright,* "but they are jobs for which a good round sum can be charged."[36]

In this context the journal believed blacksmiths could learn new lessons from Longfellow's famous verse, publishing the following variation in late 1901:

> Under a spreading blacksmith sign the village blacksmith sat;
> He heard the chuf-chuf-chuf and said: "Where is my business at?
> The road is full of horseless things, and bikes and such as that."
>
> . . . .
>
> And through his crisp and curly hair his sinewy hand he ran.
> Says he: "I'll get some different tools. As well as any man
> I'll mend a punctured rubber tire—I'll charge whate'er I can."
>
> Week in, week out, from morn 'till night, his bellows blows no fires.
> Instead it feeds a rubber tube that blows up rubber tires.
> He has a tank of gasoline, and cement, pipes and wires.
>
> And children coming home from school rubber in the open door,
> They rubber at the rubber tube a-rubbering 'round the floor,
> They rubber at the rubbersmith who rubbers tires that tore.
>
> He can't go Sunday, to the church, for that's his busy day.
> Some city chauffeur's in the lurch, and here is work and pay.
> The chauffeur buys some gasoline and chuf-chufs on his way.

But never mind, his daughter's there, up in the choir stand;
And as she holds the hymn book high shows diamonds on her hands,
For daughter's buying jewelry and dad is buying lands.

. . . .

Thanks, thanks to thee, my worthy friend, on the lesson I'll meditate,
All must at times get different tools, this world will never wait;
If we would live the strenuous life we must keep up to date.[37]

Thus, by charging whatever he could to city chauffeurs "in a lurch," the "up to date" blacksmith might put diamonds on his daughter's hand and property in his own.

Such overtly classist and regionalist arguments for smiths to take up auto work were not long-lived in the pages of *Blacksmith and Wheelwright,* perhaps because of the publicity that automotive trade publications began giving to stories of motorists feeling "gouged" by blacksmith-mechanics. Yet the fiscal advantage expected to accrue to blacksmiths taking up auto repair work remained a central theme in *Blacksmith and Wheelwright*'s sermonizing editorials. "Modern," "wide-awake," and "progressive" blacksmiths, if they followed the prescribed path, would purchase a book or correspondence course on automobile mechanisms, practice disassembling and reassembling the major components of a car when the opportunity arose, and then hang out a sign: AUTO WORK DONE HERE. The reward would be increased work and revenues and, ultimately, relief from any anxiety over being left behind by the march of progress.

For those who thought automobile mechanisms too complicated, *Blacksmith and Wheelwright* published an interview with Miss Rosalie Jones, a New York City heiress and suffragist. The anonymous writer conducted the interview with Jones "on the top floor of a busy automobile repair shop" in New York City where Jones cleaned the cylinders and ground the valves of a disassembled engine. Jones sought to master the mechanical details of motorcars so that she could break into the automobile manufacturing industry. "The automobile industry," she said, "will offer new opportunities to women. The work is not hard. It merely means using common sense." The unstated yet undeniably clear editorial message accompanying the Jones interview was that if this rich city girl can learn "all about defective carburetors, defective ignition system[s], valve trouble[s], radiator leak[s], and other ailments," then any blacksmith worth his tools should be able to also.[38]

Still, some chose not to, and Richardson and Hill invested a significant amount of editorial effort into reaching those who remained unconvinced of the necessity or inevitability of the automobile. They buttressed their editorial sermonizing with the testimonials of converts and the announcement of a "Prize Topic" contest. Prize topics on various subjects appeared long before the introduction of the automobile. Their purpose seems to have been to make the journal more practical and useful to its readers and to help keep editorial costs down by generating essentially free, original copy for publication. Typically, the journal would announce a new prize topic and offer readers cash prizes for the best articles on subjects such as "The Most Profitable Labor Saving Tool—Which Is It?" and "Is It Possible to Prevent Horses from Slipping on Wet and Icy Pavement, and if So, What Is the Best Remedy?" A given prize topic would run for a number of months with each entry published as it was received and winners announced at the conclusion of the contest.

Throughout 1902 and 1903 editorials in *Blacksmith and Wheelwright* called for readers' articles and letters describing their experience with automobiles, but with no apparent response. Then, in December 1904, the journal announced a new prize topic on the subject of repairing automobiles: "for the purpose of encouraging this new and increasing branch of the blacksmith and wheelwright's business . . . with the thought that [the prizes] will call out some practical and useful ideas, not only in relation to the repair of their various parts, but concerning the necessary equipment of tools and appliances to do the work."[39] Still, there was little response from readers. The first article on the new prize topic did not appear until June 1905, and it had a style, tone, and message suspiciously similar to the journal's editorials. No closing date was given for the contest, and it is unclear if enough entries were received to warrant awarding the prize money.

Later editorials hinted that readers were more than disinterested in auto work. Some were in fact against it.[40] To the editors' credit a sampling of this critical correspondence found its way into print and can shed light on the reluctance of some blacksmiths to take up automobile repair. Such reluctance emanated from the smith's social position in the community rather than from concerns about the technical complexities of internal combustion. One Georgia blacksmith, who employed four other men in his shop, wrote: "We are in a town of about 4000 people and in one of the best counties in Georgia. There are two other shops here. We also have two auto repair shops. They do no vehicle work, and we do no auto work here, and I think that is the right way. Our work is largely from the farmer, the saw mill and the turpentine man."[41] This Georgia smith clearly did not feel the automobile imperiled his trade or his profession. He had plenty of agricultural and other work to keep his men busy and was perfectly comfortable leaving auto repair work to others.

Nick Jacobs, a Kansas blacksmith, pointed out the potential social conflicts that a smith might encounter in his shop if he took up automobile work: "They all tell us that the auto repairing business belongs to the smith. I think it is a question of importance. Should the smith do automobile repairing? For instance, the autoist brings in his car to be repaired as quickly as possible. You can get fairly started and a farmer drives up with several plows to be sharpened or maybe a horse to be shod, and your shop force is too small to take care of the farmer and at the same time continue with your auto. If you let the farmer go elsewhere you might lose a customer."[42] Jacobs, like the smith from Georgia, looked primarily to the farmer for his work and consequently felt some loyalty toward those customers which he did not feel toward the motorist—even though other parts of his letter indicated that "a few" of the six autos in his village were owned by farmers. Jacobs also conveyed the perception that motorists and farmers were impatient with each other. The motorist wanted work done "as quickly as possible" and presumably would not favor the blacksmith stopping mid-repair to accommodate the farmer's needs. Likewise, the farmer would not wait for the blacksmith to finish repairing an automobile—often viewed as a pleasure vehicle at the time—while the tools of his trade waited in idle disrepair.

The Terry Brothers, also of Kansas, expressed similar concerns about the social tensions between the two classes of customers: "The automobile and farm work do not as a rule go very well together, for if one is doing a farm line of work, and a car drives in and you do not stop at once and work for him he generally will get sore about it. On the other hand, the farmer hates to go to a shop and when he gets there find the blacksmith down under a car and thus be obliged to wait for his job."[43] Both Kansas letter writers imply that farmers were not bothered if and when they had to wait for the blacksmith to complete prior work for other farmers, only motorists. Perhaps farmers did not mind waiting alongside other farmers at the shop door but became uncomfortable waiting alongside motorists. This could have been due to perceived differences between the importance of their own work and the triviality of the motorist's pleasure. Or it could have been due to the class and regional differences between working farmers and wealthy and/or urban motorists.[44] Whatever the cause, at least some practicing smiths noticed and commented on the social tension that accompanied the automobile into the blacksmith's shop (see fig. 9). The Terry Brothers, contrary to the editorial prescription of *Blacksmith and Wheelwright,* advised readers, "So we consider if a smith has anywhere near enough work to keep him busy he will do far better to let the auto work alone, especially if there is a garage in town."[45]

Despite the reluctance and concern expressed by some blacksmiths about automobile work, *Blacksmith and Wheelwright* continued to urge readers to make the switch, reviving the auto repair prize topic in late 1915, though with only slightly better response than the first run in 1904.[46] Significantly, L. Smith Kennewick, the third-place winner of the 1915–16 prize topic, wrote on the subject of welding a cast-iron exhaust manifold—the kind of work that John Vander Voort found himself doing for the Union Garage after 1920.[47] Dispatches to *Blacksmith and Wheelwright* from technical writer James Hobart as he visited blacksmith shops around the country indicate that Kennewick and Vander Voort were not alone in subcontracting such work for local garages. Hobart related (or re-created) part of a conversation he had with a blacksmith in Michigan: "Noting a copy of *Blacksmith and Wheelwright* on the desk, together with one or two other trade journals, the writer asked Mr. Smith if he had ever done any automobile work or intended to get into that business. . . . 'Yes, I have an auto in here once in a while, but I don't have to go after that work, just take what comes to me. There are three garages in this town and I get a good bit of work from them—crooked axles, broken springs and all that sort of thing, you know.' " Hobart pushed "Mr. Smith" to go after other, more profitable aspects of automobile service and repair, but the smith countered with the now familiar refrain: "By the time I would get at an automobile job two or three horses would come in to be shod and I would have to attend to them, so what would be the use of trying two things at the same time? . . . Yes, it might work out . . . but I don't feel like taking the risk with three other repair places in town."[48]

It is easy to see why a significant number of blacksmiths would have approached automobile work in this fashion. Straightening axles and welding metal did not tax the blacksmith's abilities nor require investment in new tools or equipment. Such heavy, heat-and-hammer work did not differ from the kinds of repairs the blacksmith did on farm equipment. Furthermore, a contract arrangement with an auto repair shop would have dissipated the social tension in the blacksmith shop on two levels. First, the motorist would likely be waiting for his or her automobile at the garage, where the bulk of the repair work was performed, not at the door of the blacksmith's shop. Second, the work that the blacksmith undertook on contact was for another tradesman, not the motorist directly, thus placing the work on the same plane of importance as that done for the farmer.

As low-priced automobiles such as the Ford Model T became available, they soon blurred the line between motorists and farmers, yet blacksmiths did not realize great increases in automobile work as a result. The advent of true mass production of the Ford Model T after 1914 and the ensuing declines in purchase price led to a dramatic increase in the number of automobiles in rural America. One would anticipate that as the blacksmith's regular clientele purchased automobiles, they would bring them to his shop for repairs. No doubt some did, as we saw with Vander Voort's example, but this development did not seem to increase the likelihood of a blacksmith making the transition to full-time auto mechanic.[49]

Blacksmiths did not generally profit from the flood of Model Ts because farmers, unlike most wealthy urban motorists, undertook many of the repairs themselves.[50] Mechanical prowess and dirty hands were widely accepted hallmarks of masculinity in rural America. Farmers' past experience with agricultural machinery, their willingness to get into a machine and "figure it out," and their propensity toward thrift served them well when their Model T bucked, coughed, or seized.[51] The Ford Motor Company facilitated such owner repairs by providing buyers with a forty-five-page instruction book that included considerable detail on how to care for the engine, drive train, and chassis.[52]

Furthermore, automobility dramatically altered the geography of repair. A farmer loading a cracked plowshare onto the bed of a horse-drawn farm wagon had a limited spatial sense of where he could go to get the needed repair work performed. The same farmer seeking repairs for his automobile—provided it was not completely disabled—had a much broader geographic sense of what constituted a "local" repair establishment. The farmer with a Model T could patronize a Ford repair shop or other garage in town that he would not have considered worth the travel time using horse-drawn transportation. Finally, by the mid-1910s, and especially by the early 1920s, enough automobile repair shops had opened in rural towns that farmers could bypass the general blacksmith shop when their automobiles needed repair. The Ford Motor Company in particular encouraged Ford owners to patronize authorized service agents for repairs. By 1912 there were thirty-five hundred Ford dealers across the United States, and that number increased to ten thousand dealers and twenty-six thousand authorized service stations by 1925.[53] So, a rural blacksmith who did not aggressively seek out automobile repair work at an early stage or who merely took what auto work came in the door soon found himself unable to compete with dealerships and repair shops in nearby towns and cities.

Numerous social and technical obstacles, therefore, lay in the path between blacksmithing and auto repair. Consequently, most established blacksmiths likely chose not to embark on that path, either retiring as blacksmiths or following the trade to its eventual demise in the 1930s. In one California community thirty-two different businesses offered automobile repair services in 1921, yet only two of them had roots in the blacksmithing trade and another having begun in the carriage-making trade. By contrast, fourteen of the community's blacksmiths remained in their trade through at least 1917.[54] The healthy agricultural economy of the prewar years seems to have kept blacksmiths busy, while entrepreneurs from diverse backgrounds staked out automobile garages, agencies, and customers in their territory.

Making the Transition

What made blacksmiths who transitioned to auto repair different from other blacksmiths, and what did they have in common with early auto mechanics who came from other trade backgrounds? The archival evidence needed to answer these questions conclusively does not exist. It appears, however, that youth was an important factor. Wiley B. Mooneyhan, a California blacksmith–turned–auto mechanic, appeared, for example, in the Riverside city directory as a blacksmith as early as 1897. His son, Walter, resided with him at the family home and shop at 330 Palm Avenue and began working in 1914 as a machinist for Theodore Crossley (owner of one of the early garages in town). Then, in 1921 the family home and shop appeared in the city directory as the site of the Palm Avenue Garage of Walter Mooneyhan, with the father listed as a blacksmith for the garage. Thus, the son appears to have been the one actually to change the family business from blacksmithing to automobile repairing.[55]

In another example in 1885 Albert F. Whitney, of Hartsville, Massachusetts, opened a blacksmith shop, where his son Raymond learned the smith's trade. One day, in about 1912, "an old Zeitz truck going through town broke down and was towed into the yard by the blacksmith shop." According to Raymond: "I would go and look at it, and one day my father asked me if I was going to fix it. I said I thought I could do the job. I looked it over quite a few times. He said that if I thought I could repair it, he would speak to the man that owned it and he believed he could get me the job. . . . I don't know how many hours I spent on that old truck. I tore the engine apart and worked for weeks and weeks, and when it was done I got paid $480." Raymond continued to work at his father's shop studying automobile books at night and working on any cars that came in during the day. The volume of Raymond's automobile work eventually surpassed that of his father's forge, and the business came to be known as the Whitney Garage.[56]

Census data supports an interpretation that the Mooneyhan and Whitney experiences were likely common among blacksmith families. Comparing available United States Census data for blacksmiths with data for workers engaged in auto-related occupations, we can see a strong youthful tendency toward the latter. Established blacksmiths did not flee the trade in large numbers at the first site of an automobile. Rather, they aged and retired in their trade, while their sons took up the new technology. It is not always possible to disaggregate the census numbers of independent blacksmiths from the numbers of blacksmiths employed by manufacturing firms such as agricultural implement makers or other machine makers, but the data does show an overall age trend for the occupation as a whole (table 3).

The numbers from 1890 and 1900 show a relatively stable age profile for blacksmiths prior to mass automobility, with roughly half of the trade falling in

TABLE 3.
Age Profile of Blacksmiths Compared with Selected Other Occupations, 1890–1920
(Percent)

Occupation	Ages		
	10–24 years old	25–44 years old	45+ years old
1890:			
All blacksmiths	16.0	50.9	32.9
All males employed in manufacturing and mechanical industries	26.5	47.8	25.3
1900:			
All blacksmiths	15.8	51.0	32.9
	10–20 years old	21–44 years old	45+ years old
1910:			
All blacksmiths	n/a	n/a	n/a
Independent blacksmiths	4.2	57.3	38.5
Garage laborers	35.1	56.4	8.5
Garage keepers and managers	3.3	80.3	16.3
Repairers in auto factories	20.8	72.3	6.8
Machinists in auto factories	17.0	74.6	8.4
	10–24 years old	25–44 years old	45+ years old
1920:			
All blacksmiths	8.3	47.7	43.8
All males employed in manufacturing and mechanical industries	21.5	50.2	28.2
Garage laborers	46.1	40.8	12.8
Garage keepers and managers	9.9	68.0	22.0
Laborers in auto factories	24.9	52.8	21.9
Semiskilled Labor in auto factories	29.1	55.3	15.4
1930:			
All blacksmiths	3.4	33.1	62.7
Independent blacksmiths	1.4	28.3	70.3
All males employed in manufacturing and mechanical industries	19.6	50.1	30.3
Mechanics in automobile repair shops	21.9	67.7	10.3

SOURCE: Data drawn from U.S. *Eleventh Census;* U.S. Bureau of the Census, *Special Reports, Occupations at the Twelfth Census,* vol. 1: 1900 (Washington, D.C.: GPO, 1904); U.S., *Thirteenth Census, 1910;* U.S. Bureau of the Census, *Fourteenth Census of the United States Taken in the Year 1920,* vol. 4: *Population, Occupations* (Washington, D.C.: GPO, 1923); and U.S. Bureau of the Census, *Fifteenth Census of the United States: 1930, Population,* vol. 5: *General Report on Occupations* (Washington, D.C.: GPO, 1933).

NOTE: Census figures from 1910 do not allow direct comparison of age groups under forty-five years with those of previous or later census returns. They have been included here as closely correlated to the other age groupings as possible.

TABLE 4.
Numbers of Blacksmith Apprentices and Garage Laborers, 1890–1920

	Blacksmith Apprentices	Garage Laborers
1890	4,244	n/a
1900	8,491	n/a
1910	2,816	4,468
1920	2,661	31,450
1930	682	66,693

SOURCE: Data drawn from U.S. *Eleventh Census;* U.S.,*Special Reports, Occupations at the Twelfth Census;* U.S., *Thirteenth Census, 1910;* U.S.,*Fourteenth Census;* and U.S., *Fifteenth Census.*

the age group of twenty-five to forty-four years old. About a third were over forty-five years old, and the remainder, a little over one-sixth of the trade, were young blacksmiths under twenty-five years old. By 1920 the trade had grown significantly older as a whole. Only 8.3 percent of the more than 195,000 practicing blacksmiths were under 25 years old, while 43.8 percent were over 45. By 1930 there was no denying that blacksmithing was a trade practiced by fathers, and even grandfathers, with 62.7 percent of the remaining trade aged forty-five and up. The trend seems even sharper among independent blacksmiths, with 70.3 percent of them reported as forty-five years or older in 1930. In 1910 the number of young men learning the blacksmith's trade had declined to 2,816, a third of what it had been just ten years before (table 4).

In contrast, automobile-related trades attracted the young. Automobile mechanics, as we think of them today, did not appear in U.S. Census figures until 1930, so prior to that year we must look at other occupations in which census enumerators were likely to report workers engaged in repairing automobiles. Young men, who in generations past may have considered apprenticing with a blacksmith, now looked for ways to get their hands on automobiles (see fig. 10). With no formal apprenticeship and few training programs available, many young men interested in automobiles sought out jobs in garages or, as we saw in the previous chapter, as chauffeurs. Young men with more experience and perhaps the financial backing of their parents, as in the Mooneyhan example, may have opened their own garages. Work in the numerous automobile factories also attracted young men and those with some experience in their father or other relative's blacksmith shop would likely find advancement open to them. All of these occupations had age profiles significantly younger than blacksmiths and younger than the average of all men employed in manufacturing and mechanical industries.

So, if youth was a trait more common among early auto mechanics than blacksmithing or any other specific occupational background, what was it about automobiles that attracted young men? Certainly, younger tradesmen did not have the long-established customer and social relationships nor the years of knowledge investment that had obstructed the transition of many established blacksmiths. For this reason they may have been more willing to heed the advice of editors such as M. T. Richardson to take up auto work. Perhaps their own fathers, recognizing the trend, pushed them into auto work and out of the family business. More likely, however, many young men went willingly and enthusiastically into automobile work because they were fascinated by the technology and because they believed that girls liked to ride in cars. Working on cars—in a garage, as a chauffeur, as a mechanic—was one way to gain access to the perceived privileges of the new technology.

Fictional literature written for young readers at the time reflected and reinforced associations between youth and automobiles. Among the many authors producing juvenile literature in the early twentieth century, Edward Stratemeyer was by far the most productive, successful, and influential. Young readers loved his books, despite the condemnations of librarians and other critics, who considered them "trashy" or warned that such stories could overstimulate boys' imaginations and "blow their brains out."[57] Stratemeyer introduced a series entitled *The Motor Boys* in 1906 which proved extremely popular and ran to twenty-two volumes, with thirty-five printings of at least five thousand copies per printing.[58] Stratemeyer portrayed the motor boys as typical upper-middle-class lads, the sons of bankers and businessmen. Honest and respectful of adults, the motor boys preferred male companionship and loved tinkering with their machines. Controlling their automobile, as well as other motorized technologies such as boats and airplanes, helped these boys to control their destinies, win the girl, and gain respect. Children's stories are one important way of establishing and extending social structures through time and space. Reading the *Motor Boys,* or any of the numerous similar tales about boys and cars, was one way that the youthful enthusiasm displayed by some blacksmiths' sons would have been reinforced and perpetuated.[59]

These are difficult speculations to prove or disprove with available evidence, but one South Carolina automobile mechanic's story, preserved by the Federal Writers' Project in 1939, illustrates the kind of enthusiasm young men brought to automobile work. As a boy, Marion Jennings of Charleston, South Carolina, completed grade school and two years of "military school" but did not want to be a schoolteacher or farmer. Instead,

> I wanted to be doing things with my hands. So I went to work in a blacksmith shop. . . . Automobiles were just coming in then, and the owners of the only two

> in our village sent them to us for repairs, so I had my first training as an automobile mechanic right in that little old blacksmith's shop.
>
> But I didn't care about being a blacksmith all my life. While I liked to tinker on cars first rate, there weren't enough of them coming in to make it exciting, so I soon quit my blacksmith job and went to work as a sewing machine salesman.[60]

Jennings went on to try a number of different jobs—as telephone repairman, hot house gardener, cabinet maker, and driver for a transfer company—before deciding to be an automobile mechanic. The transfer company he worked for purchased a "ramshackle old car. . . . It was just about falling to pieces, it had such hard usage. It was minus a windshield; minus a top; and it had a chain [drive] at the side; but it sure looked good to me." The manager of the firm offered Jennings "twenty dollars a week 'with board' to drive the car and keep it in running order," and he jumped at the chance. "Pretty soon," said Jennings, "I was having the time of my life driving drummers around in that sputtering old machine. . . . And did the gals like to go motoring!" Eventually, the car broke down completely, and Jennings's boss had him haul it to Savannah to be overhauled at a garage. Jennings remained with the car at the Savannah garage and was awestruck by the "model garage" and "all the fine tools." The garage manager let him watch and work with the mechanics in the shop and even offered him a job if he ever left the transfer company. Within a month's time Jennings had quit the transfer company and returned to Savannah, beginning his career as an automobile mechanic in earnest.

Jennings's youthful excitement about automobiles stands in marked contrast to the rather grumpy, or at least contented, conservatism and skepticism expressed by some blacksmiths. This enthusiasm about the new technology provides the common thread that runs through the letters and accounts of early automobile mechanics whether they came from blacksmithing, bicycle repairing, or machine shop work. If the lack of such enthusiasm left many rural blacksmiths to finish their days in a diminishing trade, it opened a whole new trade for others.

CHAPTER 3

Creating New Mechanics

"Get those big paying jobs in a small fraction of the time and expense of learning by the experience route," promised the Detroit YMCA Automotive School. Rather than spend years as a garage apprentice sweeping the floors and polishing brass, graduates would save "literally years of work and hundreds of dollars" in lost wages.[1] Many young Americans rushed to embrace the new technology widely touted in the media, a phenomenon much like that surrounding communication and computer technologies today, and learning the intricacies of automobiles seemed essential preparation for the future. Acquiring technical knowledge about automobiles, however, remained random and haphazard. One could, as we have seen, pick up a smattering of knowledge from various published sources or from tinkering with whatever model was available. Yet this ad hoc method of learning did not satisfy all. Many Americans sought more formal and structured means for learning about the new technology. The auto school opened by New York's West Side YMCA to train chauffeur-mechanics and wealthy motorists in 1904 launched one of the first formal programs geared specifically toward teaching students how to maintain and repair automotive technology.[2] In hindsight its opening marked the beginning of a two-decade-long process of building new social and institutional structures for producing and reproducing generations of automobile mechanics. That process would lead from a multiplicity of courses and schools in the 1910s, through the massive efforts to train mechanics during World War I to the introduction of automobile repair courses in public high schools in the 1920s.

When the automobile appeared on the American scene at the turn of the twentieth century, the traditional system of training skilled workers through apprenticeships with master craftsmen had long since passed, and new systems for training workers remained in a state of flux and disarray. Never as firmly established in eighteenth-century America as it had been in Europe, the traditional apprenticeship system declined dramatically in the nineteenth century.

Industrialization, the ideology of independence, individualistic religious sentiment, and the rise of consumerism all militated against traditional, indentured apprenticeships. By the late nineteenth century very few tradesmen and virtually no factories supervised traditional, indentured apprentices. Many still used the name, but late-nineteenth-century apprenticeships mostly exploited young workers in low-wage jobs without providing any formal trade training, guidance, or opportunity for advancement.[3]

At the dawn of the twentieth century no effective alternative to traditional apprenticeships had yet evolved. In the void established industries such as the machine tool and foundry industries experimented with modified apprenticeships, employer-supported trade schools, and union-controlled training programs. Workers seeking to enter the many new occupations created by industrialization, such as bookkeeping or typing, could enroll in the growing number of commercial schools if they were fortunate enough to have money for tuition or to live in a city that sponsored such a school.[4] Most workers seemed to learn, however, by moving from job to job, picking up what knowledge they could as they went along. Repairing automobiles emerged as one of those new occupations with no tradition of apprenticeship and no effective alternative.

Increasing demand for auto mechanics led urban reformers, entrepreneurs, and even the U.S. Army to establish schools and training programs for creating new automobile mechanics, and eager students enrolled in them. The quality of these schools and their programs varied widely, yet the directors and proprietors all hoped to profit from the public's desire to learn more about the new machines. With nowhere else to turn, those interested in learning about the new technology by more than trial-and-error or tinkering methods turned to the new auto schools and correspondence schools for help.

Learning about Automobiles at the YMCA

The seemingly unlikely role that the YMCA played in creating new mechanics emerged from the organization's Christian urban reform mission. Saving young men's souls included helping them integrate into society through education and appropriate employment. After an earlier focus on educating young men for the ministry, the International Convention of the YMCA endorsed general educational work in 1889 "as a function of the Association."[5] Shortly thereafter, the YMCA embraced vocational preparation and training as part of its educational mission. By 1900, 288 chapters nationwide enrolled over 24,000 students in courses ranging from English and mathematics to commercial art and mechanical drawing.[6] Among the early chapters to pursue this expanded educational mission aggressively, the New York YMCA enrolled 1,747 students at 7 branches during the 1901–2 school year.[7] The West Side and the Twenty-third Street branches enrolled the majority of these students—in traditional academic courses as well as commercial and trade classes.

A 1902 study by Walter Hervey, chairman of the West Side education committee, accelerated the branch's movement into vocational curriculum by recommending additional vocational courses "adapted to the multiple needs of employed men" in New York's Manhattan district.[8] Accordingly, the West Side's brochure for 1903–4 offered prospective students a menu of forty-nine classes ranging from geometry and trigonometry to bookkeeping, commercial law, architectural drawing, mechanical drawing, carriage drafting, penmanship, and typing. By the 1904–5 school year 1,156 men enrolled in classes at the West Side, and tuition receipts from January through November 1905 totaled over twenty-nine thousand dollars, representing "a growth of over 2000% in five years."[9]

In this climate of an expanding educational mission the Automobile Club of America (ACA) easily persuaded the education committee of the West Side to undertake the training of chauffeur-mechanics. Such a course of action seemed ideally suited the Y's twofold objective of practical job training and Christian character building. In concert with the ACA's wishes, the YMCA directors sought to produce "competent and reliable men" who would be employed by the city's wealthy motorists and garage managers.[10] With the moral self-assurance of Progressive Era reformers, their education committee established a school that would both train men to fill moderately well-paying positions and help remedy the social evil of the chauffeur problem by instilling proper work habits, temperance, and manners in its chauffeur-mechanic graduates.

The education committee must also have sensed the enormous revenue-generating potential that an auto school represented in terms of course fees and new association memberships. By the end of its first year of operation 365 students had taken one or more of the three auto classes offered, and receipts for the 1905 calendar year totaled nearly twenty thousand dollars, swamping the income generated by all the other courses at the West Side. The ACA continued to donate to the auto school, and various automobile manufacturers and dealers provided parts, equipment, and even whole cars. Favorable news coverage in the city papers and in the automotive trade press helped the West Side's auto school grow steadily for nearly two decades. Throughout the pre–World War I period auto school enrollments consistently outnumbered those of all other courses offered at the West Side, crossing the 1,000-student annual enrollment level in 1909–10 and the 2,000-student mark by 1916–17 (see fig. 11).[11] While the YMCA was technically nonprofit, the revenues of the auto school paid the salaries of many instructors, helpers, and staff as well as a significant portion of the overhead for two of the West Side's buildings. Auto school students also became a regular source of new dues-paying members for the West Side YMCA. Thus, strong institutional incentives soon developed to maintain and enlarge the West Side auto school.

The early West Side auto school courses appealed essentially to two populations: wealthy and middle-class motorists who wanted to learn more about their own machines; and aspiring chauffeurs, garage helpers, and would-be mechanics who wanted to improve their situation in life by learning how to drive, maintain, and repair the new technology. In order to reach these two groups, the West Side initially offered three automobile courses: an illustrated lecture series on automobile types (steam, electric, and gasoline) and their principles of operation; a design and drafting class for "men who desire to study the general theory and practice in the design of motors and their accessories"; and a class entitled "operative work" that encompassed hands-on garage work and "road work" in which students learned to drive and maintain different types of automobiles. Columbia University engineering professors taught the first two courses, while Clarence B. Brokaw, described by the school as "an authority on automobile matters," supervised the third. By the 1905–6 school year the design and drafting course as well as the Columbia professors had been dropped, and Brokaw, now principal of the auto school, delivered the lectures on automobile principles and supervised a teaching staff for separate "garage laboratory" classes and road work classes.[12] The auto school offered these three classes essentially unchanged until a major revamping of the program in 1919.

The West Side's auto classes varied in length over the years but were surprisingly short in comparison to traditional skilled apprenticeships. The lecture class met evenings once a week for three months; the garage laboratory or shop class met three hours per week for three months either mornings, afternoons, or evenings; and the road work class took about two and a half months to complete, being scheduled around the weather and "the condition of the cars." By 1909–10 the West Side auto school had reduced road work to eight one-hour lessons spread over four to eight weeks and compressed the shop class into a four- to eight-week course composed of sixteen three-hour sessions. Responding to technological developments, the education committee added a course on self-starters and ignition systems in 1914. This cost slightly less than the garage laboratory course and thus presumably ran somewhat less than thirty-six total instructional hours. In the context of past multiyear apprenticeships for skilled machinists, these courses seemed brief to some, and critics doubted that students desiring to become mechanics could learn sufficient skills in such a short time.[13] The Y's goal, however, was merely to give such men the requisite technical training to begin them on a productive path in the automobile service field.

While more than half of the students who enrolled in the auto school during the first ten years trained as chauffeurs or mechanics, the remainder sought instruction as motorists or were considering purchasing an automobile.[14] This dual curriculum for owners and chauffeur-mechanics served the auto school well during its early years but delayed the development of a more extensive auto repair course. The constant scramble to keep the classroom engines, chassis, carburetors, and other parts up-to-date with rapidly evolving automobile technology led school personnel regularly to solicit manufacturers and dealers for donations. The education committee noted in 1916 that having students "of the owner type . . . is the best argument in securing equipment."[15] Manufacturers and dealers would donate cars and car parts in the belief that new motorists who learned about automobile technology using their company's equipment would leave the school with a strong brand loyalty. This dependence upon manufacturers for basic classroom equipment led the YMCA to emphasize and even depend upon enrolling wealthy motorists and potential motorists in order to maintain relatively current shop equipment. Teaching wealthy motorists how to manage their machines and supervise their chauffeurs, however, did not get to the heart of the YMCA's educational mission.

Fiscal incentives may have kept the education committee interested in wealthy student motorists, but the perceived needs of the student mechanic engaged their moral and social sense. They emphasized "character building" among these students as equally important with technical training. Representatives of the religious work department held regular meetings in the automobile shop and addressed each new group of students as they entered the auto school. Likewise, auto school instructors attended regular Wednesday night Bible classes. Beyond purely evangelical purposes the character-building emphasis seemed aimed at reducing tensions between worker-graduates and their employers by instilling the idea of "Christian service" in students (i.e., teaching them to respect their employers and to strive to help others) and training students in what would be expected of them in the workplace. The executive secretary of the United YMCA Schools asserted in 1923 that the "distinctive objective" of YMCA educational work "is the development of Christian ideals, attitudes, and habits in its students." This was perhaps as important to prospective employers as was the brief technical training imparted by the YMCA courses.[16]

Segregation at the Y

Throughout the West Side auto school's early years the education committee practiced various forms of social segregation which both reflected and reinforced the emerging social identity of automobile mechanics. The school's first brochure carefully pointed out that it taught the operative course separately, with one section for automobile owners and another for chauffeurs. Essentially segregation by social class, this system reflected the social distinctions and tensions that fueled the chauffeur problem and inspired the formation of the school in the first place. As the auto school became more popular, the broad appeal of automobility also forced the education committee to confront other forms of social segregation common in American society at the time.

Despite their absence from traditional histories of America's automobile culture, African Americans shared an interest in automobiles. When the opportunity arose, blacks embraced automobile consumption and travel as quickly and enthusiastically as any Americans in the early twentieth century—perhaps more so.[17] Blacks' interest in the freedom of automobility was no doubt aided in the early 1900s as numerous southern cities adopted ordinances segregating their street rail lines. In more than two dozen such cities blacks responded by boycotting the lines and seeking alternative transportation or just walking. Reflecting Booker T. Washington's influential strategies of building up black business and economic resources while avoiding direct confrontation with whites, prominent Nashville leader Richard Henry Boyd told the 1903 meeting of the National Negro Business League: "These discriminations are only blessings in disguise. They stimulate and encourage rather than cower and humiliate the true, ambitious, self-determined Negro."[18] When Nashville segregated its streetcars in 1905, Boyd portrayed it as an opportunity for "stimulating the cause of the automobile as a common carrier."[19] He joined other black city leaders to form the Union Transportation Company, which invested ten thousand dollars in fourteen electric buses and aimed at both making a profit and getting around "the nefarious law of Jim Crow street cars."[20] In the hostile legal climate of the time, most of these boycotts failed in ending segregation, and in the face of the political and economic advantages enjoyed by the streetcar companies, alternative black-owned transportation companies could not turn a profit.

By the early 1920s numerous black entrepreneurs instead operated small "jitney buses" in southern cities as an alternative to segregated streetcars. The small, five- to seven-passenger jitneys required a lower capital investment than Boyd's earlier experiment, and they offered more direct service to black and working-class neighborhoods for those who could not afford a private car. Their success and proliferation in some cities led streetcar companies and white newspaper editors to call for and eventually secure licensing and registration fee legislation effectively banning jitneys by the mid-1920s.[21] In this context of increasing Jim Crow segregation, the freedom of automobility meant perhaps more to black Americans than it could have to almost any white American.

Not all blacks purchased automobiles as expressions of protest. Many bought them for the same reasons white men and women did—novelty, utility, and status. Automobiles offered escape from urban congestion for some or and end to rural isolation for others. Middle-class blacks also used the status of automobile ownership and travel to support a racial uplift ideology in the 1920s and 1930s.[22] For poor and working-class blacks purchasing a car may have been less of a concern than learning how to drive and work on one in order to secure a comparatively prestigious job as a chauffeur or a mechanic.

This was likely the motivation that brought the first black student to the door of the West Side YMCA in 1908 and pushed the education committee to consider allowing "colored men" to enroll in the auto school. They determined they could do so only if enough black students enrolled to open a separate, segregated class.[23] There is no indication that the West Side ever opened a segregated automobile class, but they appear to have investigated where in town they could refer inquiries.[24]

Blacks denied entry at the West Side YMCA's auto school may have enrolled in the Cosmopolitan Automobile School on West Fifty-third Street. There manager Lee A. Pollard offered to teach students enough "of the theory and practice of automobile[s] and automobiling to enable them to meet the emergencies that constantly arise" on the road. Cosmopolitan offered what was clearly a chauffeur's course developed during the tumultuous days of the chauffeur problem in New York, yet students might have hoped to use its courses as a means to enter various aspects of automobile work. Pollard's school also provides an early example of African Americans beginning to develop their own network of automotive knowledge and services. It is not clear how long the school survived or how many students it graduated.[25] What is clear is that racial segregation was becoming institutionalized in automobile service in ways that would grow increasingly pronounced after World War I.

Women also wanted to learn more about the new technology, contrary to the gender stereotypes of the early twentieth century, and soon sought admission to the West Side's auto school. In the early twentieth century numerous upper-class women such as Edith Wharton, Alice Roosevelt—daughter of the president—and an army of women suffragists drove their own automobiles in symbolic rejection of the cloistered female sphere of family and home. At the same time, female race car drivers such as Joan Newton Cuneo achieved national fame before they were banned from the sport in 1909.[26] Most early female motoring seems limited to wealthy society women who could afford their own cars and mechanical tutors, but women of lesser means might have also sought to get their hands on the wheel and the wrench. We simply do not have their writings and reflections with which to verify or explore their experiences. What is certain is that the West Side's education committee finally recognized the swelling demand among Manhattan women for automobile instruction and opened a gender-segregated road work course for women in 1913. Even so, women's presence in the auto school remained tenuous, as they were the first to be bumped to accommodate male enrollments.[27]

While limiting or denying access to certain groups, the West Side welcomed and encouraged out-of-town students to live at the Y and study in the auto school. They offered dormitory rooms in the YMCA building and maintained a directory of "reliable" rooms and boardinghouses in the Manhattan area. A December 1920 study of 800 auto students revealed that 167 came from out of town. In March 1921 a study of all West Side students showed that 292 of the 363 out-of-town students enrolled in automobile courses. Most resided in neighboring New York counties, Long Island, or New Jersey, though 84 of them arrived from other states, and one came all the way from Mexico.[28]

The West Side's auto school grew in influence as well as size during its first two decades. In addition to swelling enrollments that by 1917 filled classes in two buildings—one on Sixty-sixth Street and the other on Fifty-seventh Street—the school enjoyed a high profile in the emerging field of automotive trade education. The International Motor Company of New York sought out the West Side's help in training their truck drivers and mechanics in 1912,[29] and in 1916 the Bronx branch of the YMCA consulted with the West Side before establishing its own auto school. In the same year Mr. Drum, West Side's chief mechanic, left to become principal of the Bedford YMCA auto school.[30] Harry C. Brokaw joined Clarence B. Brokaw, principal of the West Side auto school, on staff and eventually became the West Side's technical director in 1918.[31] H. C. Brokaw traveled regularly to investigate and consult with other private and YMCA auto schools, and in 1910 he coauthored a popular automobile textbook.[32] In 1919 he became an active member of National Automobile Dealers Association and in 1920 joined the Society of Automotive Engineers, with the West Side auto school picking up the cost of his dues.[33]

The educational work of the West Side in general and the auto school in particular gained national influence when the YMCA international committee convened a meeting in Detroit in 1919 to begin making plans for standardizing course curricula among YMCA branches nationwide. Walter Hervey, chair of the West Side's education committee, gained appointment to the executive council charged with selecting and overseeing an International Educational Staff.[34] Soon H. C. Brokaw began serving on the commission writing the standard automobile syllabus, and by 1920 the national YMCA organization urged all Y-sponsored auto schools to "conform with the standardized program."[35] By the end of the 1920 academic year at least sixty YMCA organizations conducted complete auto schools, and fifteen others offered small classes or lectures on automobiles, enrolling a total of approximately 13,500 students that year.[36]

Other Early Auto Schools

The early success of the West Side's auto school did not go unchallenged. The number of profit-oriented commercial automobile schools multiplied rapidly during the decade before America's entry into World War I. These schools reflected their founders' entrepreneurial spirit as well as the large unmet demand for automotive training. Young men, and many women, eagerly filled their seats and bought their correspondence materials.

Early commercial schools ranged in quality as each tried to profit in its own way from the new technology. Like the West Side YMCA, many began by teaching chauffeur-mechanics and motorists how to operate and care for cars. They, too, offered short, four- to twelve-week courses that could have imparted no more than a cursory knowledge of automobile maintenance and repair. Proprietors of such schools enjoyed relatively unfettered freedom to meet market demands for automobile technical education. Neither automobile makers, automobile dealers, nor labor unions directly controlled these early auto schools.[37] Nor did commercial operators need to abide by any government-imposed educational guidelines. As a result, some schools conducted outright scams, such as C. A. Coey's School of Motoring in Chicago, which was more concerned with securing sales agents for the Coey Flyer automobile than with producing competent chauffeurs and mechanics through its fifteen-dollar correspondence course.[38] The Practical Auto School, operating out of the Coffee Exchange building in New York, offered students two-dollar commissions on sales of its "Home Study Course" to friends. "If you are ready to help us," the school's 1912 brochure promised, "you should be able to make from $8 to $15 a week. If you can devote more time to soliciting for us, your earnings will be proportionally larger."[39] Some of these early commercial auto schools no doubt made quick profits by promoting their commission rates rather than their completion rates.

Other, larger schools put more effort into their curriculum. The Sweeney Automobile and Tractor School and the Rahe Auto and Tractor School, both in Kansas City, Missouri, and the Michigan State Automobile School in Detroit grew through reputation and skillful marketing to rival and eventually surpass the enrollment figures of the West Side YMCA's auto school.[40] By 1920 the national YMCA organization estimated that commercial auto schools nationwide enrolled about twice as many students as the YMCA auto schools, or about twenty-seven thousand annually.[41]

The perception that automobile schools had "sprung up overnight like mushrooms" prompted criticism, journalistic investigations, and calls for their regulation—largely from those hiring their graduates. In this context the YMCA auto schools always received praise for being backed by trustworthy organizations, even if their equipment and facilities were not always on par with the better private schools. Despite much editorial consternation on the matter, little was done to control the new auto school entrepreneurs. Rather, students voted with their dollars for the schools that fulfilled their expectations, thus helping sift the more flagrant frauds from the field.[42]

What, exactly, did these students expect from auto school courses? Working-class students used the courses offered by the International Correspondence School (ICS) in Scranton, Pennsylvania, to improve their income and, more important, to move to "the other side of the desk"—that is, to move into white-collar work. The courses offered by ICS ranged from "English Language for Foreigners" to commercial law and structural engineering. Clearly, completion of some of ICS's courses of study could qualify graduates to move to the other side of the desk and leave dirty, toilsome work behind.[43] Yet what of the students who took ICS's course in automobile mechanics or those who took the automobile courses offered by the American School of Correspondence or who enrolled in any of the dozens of other resident or correspondence automobile schools—what did they hope to gain by learning about carburetors and ignition systems? Were they simply interested in things mechanical, or did they believe that such an education would improve their income or advance their class standing in their community?

In strictly economic terms, as long as the demand for chauffeurs and garage mechanics exceeded the supply, the resulting high wages would attract more to the field and thus to the schools. The schools certainly appealed to and promoted prospective students' desire to earn more money. School promoters intentionally conflated their curriculum with the growth of the automobile industry generally, vaguely associating the training they offered with the riches of Detroit's captains of industry. The Portland Auto School claimed in 1907 that "the Automobile

Business offers the young men of this age one of the best opportunities . . . to get into a profitable occupation. . . . It is one of the few lines of business in the country that gives the poor man a chance to get ahead in life and make some money" (see fig. 12).[44] The National Auto School in Cincinnati assured prospective students that "many men trained in this school enjoy the prosperity which is sure to come to one who prepares himself properly to enter a growing, booming field like the auto business."[45]

The potential physical freedom offered by automobile work provided another enticement to auto school students. Many auto schools dangled alluring visions before their prospects. The Practical Auto School described the benefits of becoming a chauffeur-mechanic in especially attractive terms in 1912:

> Spring follows closely upon the heels of Winter, opening up new delights to the chauffeur—delights that no other mode of transportation can equal. Drives out of town become more frequent as the budding leaves and flowers lure you further each week end.
>
> In fact every season has its individual delights, but perhaps the most enjoyable is summer. Then it is no hardship to tumble out of bed early and start the day with a cool invigorating spin. The joy of this early ride is indescribable, literally flying over the country, leaving behind dull cares and starting the day fresh in body and clear in mind. . . .
>
> Remember, too, the pleasant, unexpected meeting with other jolly chauffeurs as you roll up for lunch at some fine hotel or wayside inn. What motoring experiences you compare over cards and cigars![46]

Ignoring the frustrations of blowouts, dirty repair jobs, and demanding employers, Practical Auto School offered working-class, cards-and-cigars-type young men a vision of physical freedom which they could easily contrast with factory or indoor work that might await them. The Automobile College of Washington, D.C., reversed the other-side-of-the-desk imagery employed by the International Correspondence School by rhetorically asking readers of its catalog to choose between "$10.00 per week, long hours, and drudgery" as a clerk working indoors under a glaring incandescent lightbulb or "$30.00 per week and short hours in the open air" as a chauffeur-mechanic.

Auto schools also tapped into one of the enduring attractions of the mechanic's trade by selling the idea of masculine occupational independence. The Michigan State Auto School stated flatly in its recurring advertisements, "Get a Better Job—or—Go into Business." In an updated version of Thomas Jefferson's yeoman farmer or Longfellow's independent village blacksmith, E. J. Sweeney, president of the Sweeney auto school, promised: "I not only teach you this business, but I try and endeavor to teach you how to make money at it. . . . I try to make every man go into business for himself . . . and be his own boss."[47] As the Practical Auto School claimed, few other occupations seemed so easy to enter and quickly qualify for independent proprietorship—one could become an independent tradesman without enduring a long apprenticeship. Auto repair seemed to appeal, in part, to an aspiring petit bourgeois. Even if one worked for a large garage or a demanding employer, the potential always existed to hang out a shingle and open one's own repair shop. This helps explain difficulties that labor unions later experienced in organizing automobile mechanics.[48] Not only were mechanics scattered among thousands of small shops; many did not see themselves as permanently working-class, instead nursing visions of the day they would be on the "other side of the desk" as proprietors and possibly even employers.

By the mid-1910s the field of automotive technical training remained a jumble of commercial schools and YMCA schools, of correspondence courses and get-rich-quick scams, of traveling factory reps and how-to books, trade journals, and manuals. This messy array nonetheless offered multiple avenues of entry into the nascent occupation. Interested women as well as men, motorists as well as mechanics, enrolled in courses and learned about automotive technology. The variety of short-course auto schools and correspondence programs offered the formal means for these aspirants to gain automotive technical knowledge prior to World War I. Some observers began to fear that such a mishmash system of training would not meet the needs of a growing automotive consumer market. It was not clear, however, just how the mix would, or should, shake out.

Viewed in retrospect, the records of the West Side YMCA's auto school reveal a trend that may have been typical for other auto schools.[49] Student enrollments in the West Side YMCA had reached a plateau and began to decline in 1916.[50] Given time to ponder this trend, the education committee might have been able to see that its dual-focus curriculum—for owners and for mechanics—was becoming outdated. Wealthy Manhattan motorists had become slightly more comfortable with the now twenty-year-old technology, and in the future they would not flock to the YMCA courses as they once had. In addition, the auto mechanic's trade was becoming more clearly defined, and to some in the growing industry the short courses offered by the YMCA seemed increasingly inadequate preparation, even for novice mechanics.

By implication commercial and YMCA-sponsored auto schools nationwide may have felt similar effects of maturation and might have taken steps to adjust in different ways. Even so, in 1917 the fates of these schools and of tens of thousands of young American mechanics and would-be mechanics soon became swept up in war. As America geared up for mechanized war, training truck drivers and mechanics quickly moved to the top of the military's priority list. The West Side YMCA and hundreds of other automotive and technical schools experienced record-high enrollments by rallying to the service of what one contemporary observer called the "University of Uncle Sam."[51]

The University of Uncle Sam

When President Woodrow Wilson asked Congress for a declaration of war in April 1917, the United States Army stood woefully unprepared for modern, mechanized warfare. Years of sparse peacetime budgets and entrenched old-army attitudes ensured that the U.S. Army lagged behind the armies of Europe—particularly in motorization. The French army had begun experimenting with motor vehicles in military maneuvers as early as 1897 and, along with Germany and Great Britain, began offering subsidies to private auto and truck purchasers on the contingency that they would lend their vehicles to the military in case of war. Thus, the French army had immediate access to seven thousand trucks at the outbreak of the war and could quickly requisition seventy thousand to eighty thousand more vehicles.[52]

The United States, by contrast, had most recently emerged from the Spanish-American War, which highlighted the importance of naval power. Consequently, the U.S. Army had conducted only a few, very limited studies of the military use of automobiles and trucks before the outbreak of war in Europe. The studies never proved to America's top brass that automotive technology could be more cost-effective than the familiar railroad lines and mule trains. Motor trucks and their requisite maintenance and repair equipment remained relatively expensive, and congressional money was not forthcoming. In 1915 the U.S. Army possessed only twenty-seven automobiles and few trucks.[53] Then on 9 March 1916 Pancho Villa and his forces surprised the U.S. Cavalry at Columbus, New Mexico, and initiated what Americans called the Mexican Punitive Expedition.

The Mexican Expedition—a running land battle in an area not well served by rail lines—provided the impetus to initiate a full-scale test of army motorization. Within a week of Villa's raid two motor truck companies—one composed of 27 vehicles from the White Motor Truck Company and the other made up of 27 Jeffrey Quads—joined Pershing's foray into Mexico. By the end of June the Quartermaster Corps of the U.S. Army had purchased more than 588 motor trucks, 75 automobiles, 61 motorcycles, 10 machine shop trucks, and 6 tow trucks as well as fuel trucks, road-building tractors, and an inventory of spare parts and supplies.

The use, maintenance, and repair of these vehicles provided the U.S. Army with its first real experience in motorization and highlighted the need for standardization—in equipment and personnel. After trying to coordinate parts and service for thirteen types of trucks from eight different manufacturers, army planners realized the value of standardization and began working with the Society of Automotive Engineers to develop the "Standard B," or "Liberty," truck. Meanwhile, Maj. Francis H. Pope drew on his experience in charge of the repair shops at Fort Sam Houston to help formulate army doctrine on things such as the number of trucks in a "truck company," the rules of convoying, and the number of mechanics required to keep a given number of trucks operational.[54]

Procuring competent mechanical personnel for a motorized army during full mobilization proved problematic. The commercial firms supplying the trucks had assembled the motor truck companies of Pershing's expedition. In the case of full mobilization the demand for skilled drivers and mechanics by both industry and the army quickly would exceed the nation's supply.[55] Army planners were still studying the lessons of the Mexican Expedition when events in the Atlantic pushed President Wilson to seek a declaration of war against Germany.

The first emphasis in America's war strategy was to get fighting troops to France as soon as possible, but the demand for motor transportation soon grew critical.[56] In May 1917 the army brass brought Capt. (later Col.) Harry A. Hegeman to Washington, D.C., to organize and equip a Reconstruction Park for the American Expeditionary Forces in France.[57] Hegeman, a mechanical engineer by training, had commanded Motor Truck Company No. 1 under General Pershing in Mexico and had organized and equipped the general repair shops at Fort Sam Houston. Upon arriving in Washington, he recruited officers for his new organization through the personal contacts in industry he had established during his Mexican experience. He had trouble, however, getting the requisite number of skilled enlisted men—a problem faced by the army in general. Hegeman tried using lists of skilled men of draft age provided by employers, he tried transferring skilled men from other organizations within the army, and he tried drawing recruits from the general draft pool, all with a very low success rate. Employers did not want to part with their experienced mechanics, army commanders bureaucratically "hid" their skilled men from anyone who would transfer them, and the skill level of the general draft pool was too low to fill Hegeman's needs.[58]

Hegeman finally received authorization for himself and his handpicked officers to go to manufacturing cities such as Detroit, Cleveland, Akron, Toledo, and Philadelphia. There they made patriotic appeals at the gates of the factories, encouraging skilled men to join Hegeman's organization. Hegeman eventually gathered the thirty-four hundred men he needed to form three Motor Repair Units. These units set sail for France in January 1918 and eventually became the core personnel of the Reconstruction Park at Verneuil, France. Although Hegeman had found the men he needed, army planners realized that such recruiting methods would be insufficient over the long run. Officers could not recruit indefinitely at the factory gates without soon jeopardizing industry's ability to produce materials for the war.

Army planners approached the problem of recruiting for a modern, mechanized army from two directions: by better organizing the technical men within their ranks and by training tens of thousands more men in the mechanical details of modern war machines. The Committee on the Classification of Personnel (CCP), created in February 1918, attempted to identify and quantify the personnel needs of the army.[59] The CCP worked closely with the various branches of the army to develop a classification system for the technical specialists needed by the military and the civilian occupations which most closely corresponded to each of those needs.[60] The committee further developed a series of oral, written, and performance trade tests to evaluate each recruit's depth of knowledge and experience in a given trade.[61] Using the classification system and the trade tests, in conjunction with the more widely known Alpha and Beta intelligence tests, the army sought to coordinate and rationalize its personnel system.[62] The classification and testing system may have increased the efficiency with which the army assigned the men it received, but it did not alleviate the general shortage of men with the desired technological knowledge. Alleviating that shortage meant embarking on a massive training program.[63]

The army created more truck drivers, mechanics, and motor corps officers by establishing Motor Transportation Schools at Camp Holabird in Baltimore, Camp Jessup in Georgia, and Forts Bliss and Sam Houston in Texas.[64] As the army geared up to train mechanics at these camps, however, planners projected that as many as a half-million soldier-mechanics would be needed to achieve full mobilization. They realized that the army could not train enough mechanics and technicians—in motor transport, telegraphy, aviation, and myriad other trades—solely at government facilities in time to meet the nation's needs. Thus, the secretary of war created the Committee on Education and Special Training (CEST) in February 1918. This committee coordinated the training of soldiers as technical specialists through a network of private and public trade schools and colleges around the country. By the end of August 1918 forty-seven thousand men, scattered among more than three hundred schools, studied various technical trades such as blacksmithing, machining, aviation, and motor repair.[65]

The wartime diary of David McNeal provides a rare insight into the transformation of an inexperienced enlistee into an army mechanic via the motor transportation school at Camp Holabird in Baltimore, Maryland. McNeal, of Davenport, Washington, signed up for the army at the enlistment office in Spokane on Thursday, 27 June 1918.[66] McNeal, ostensibly eighteen years old but still attending high school, needed his parents' signature on his enlistment papers. Reporting to Fort Wright the next day, he spent the weekend undergoing physical exams and bureaucratic processing. On Monday morning the army put him on a train for the six-day ride to Camp Holabird. Once there he began an intensive eight-week course in the school of motor transportation. David McNeal, now Private McNeal, had apparently never driven a truck before, but the army desperately needed drivers and mechanics, both at home and in France.

McNeal's diary chronicles the daily combination of lectures, hands-on shop work, and self-study which was typical of the army's technical courses. On his first day he heard a lecture on trucks. On the morning of the second day he began learning how to drive a truck. At a time when automobile makers were just beginning to make passenger cars more comfortable and user-friendly, truck makers still battled basic reliability problems and devoted relatively few resources toward easing demands on the truck driver. Simply maneuvering the army's huge trucks required great shoulder and arm strength, and operating their nonsynchronized manual transmissions took much practice. By the fall of 1918 the army's Committee on the Classification of Personnel developed a trade test for truck drivers which required maneuvering a truck through a sinuous path of wooden posts placed nine feet apart—without hitting the posts.[67]

The afternoon of McNeal's second day at Camp Holabird his class got down to business and began overhauling a White Motor truck. In the following days and weeks Private McNeal learned how to grind and adjust valves; tighten bearings and connecting rods; and disassemble, reassemble, rebuild, and repair magnetos, carburetors, clutches, u-joints, transmissions, rear axles, and more. Many of his evenings were devoted to "trouble shooting" motor problems concocted by his instructors. Private McNeal's class also heard lectures about each of the major subassemblies of motor vehicles as well as lectures from a "Goodyear man" on the construction and repair of tires and from a "Cpt. Evans" on his experience with motor trucks in the Mexican Expedition.

Toward the end of the course the army packed Private McNeal and his classmates off to Detroit, where they toured the Dodge and Ford factories prior to receiving a shipment of new trucks and parts destined for the American forces in France. Eastbound shipping space on railroads came at a premium during the war, so the army sent soldiers to drive motor truck deliveries from the factories to the coastal embarkment centers. McNeal and his cohorts convoyed their vehicles, laden with spare parts, overland to Baltimore, and upon arriving successfully at Camp Holabird, they graduated from the mechanics' course.[68] Private McNeal received a certificate from the school verifying his proficiency in the operation, maintenance, care, and repair of motor vehicles and shortly thereafter set sail for France.[69] While in France, the army kept Private McNeal busy working on Fords, GMC trucks, Indian motorcycles, even Pierce Arrows, until his discharge in August 1919, at which point he concluded his diary.[70]

Other soldiers had similar experiences as they underwent training at the cooperating private and public technical schools around the country. An engineering journal of the time profiled the University of Pittsburgh's program for producing "fighting mechanics" for the army. There the university provided housing, meals, and instruction for fifteen hundred student-soldiers in return for a per capita / per diem fee paid by the army. The university constructed seven two-story barracks, a large mess hall, and a 16,000 square foot machine shop to accommodate the fighting mechanics.[71] Instruction consisted of a combination of lectures and hands-on shop work, much the same as Private McNeal experienced at Camp Holabird. For each of the eight weeks of training at the University of Pittsburgh, groups of twenty student-soldiers learned about a particular aspect of motor vehicle repair such as engine overhauling, magneto and ignition timing, or spring and chassis work.

Not surprisingly, Private McNeal and his cohorts in the various training centers learned about motor vehicles in racially segregated settings. Since the U.S. military was not integrated until after World War II, the army needed black truck drivers and mechanics to support its segregated units. The CEST therefore contracted with a dozen black schools to undertake the training of black soldiers in various technical trades.

The army did not seek to undermine or challenge white stereotypes of blacks as mechanically inept, unsophisticated, or "backwards." The attitudes of the white directors, officers, and civilians involved in training black soldiers in technical trades no doubt limited the quality and extent of such training. The educational director for the CEST's third district dismissed the training conducted at Howard College (now Howard University) in Washington, D.C., by writing in his final report of 1918, "The job of teaching trade work to these colored men was a difficult task because of racial characteristics."[72] Such attitudes also help explain why, of the six black schools identified in the CEST final reports as conducting motor vehicle training, three of them offered only driving courses, two of the three that offered mechanics training were described as "weak," and only one, Wilberforce, in Ohio, received an "outstanding" evaluation from its white CEST director.[73] Despite these limitations on black technical training, an unknown number of young black men gained hands-on experience with motor vehicles in the University of Uncle Sam, and a handful of important black schools gained experience and equipment for propagating technological knowledge after the war's conclusion.

The army largely succeeded in filling the seats in all of the cooperating schools by making regular public appeals for recruits. Press releases, patriotic appeals from General Pershing, and even full-page ads in the *New York Times* encouraged men of draft age to volunteer to join the "Gas Hounds" (see fig. 13).[74] Although the army never achieved full mobilization prior to the Armistice, the number of army personnel involved in motor transport increased from essentially zero in March 1917 to 103,000 by November 1918, with more than 37,000 serving overseas.[75]

Herschel Hunt was one of the many who heeded the army's call. Born in Page, Nebraska, he enlisted in the army at O'Neill, Nebraska, on 13 August 1918. Hunt listed his previous occupation as "farmer" but later recalled, "when the Army found that I had former experience in auto mechanics they sent me for a month to a Auto and Tractor School which I found very agreeable and educational in Kansas City." Hunt's discharge papers indicate that the army sent him to the Rahe Auto and Tractor School in Kansas City, Missouri, from 13 August to 16 October 1918. After that he was assigned to Motor Transport Repair Unit 316, Camp Boyd, El Paso, Texas, for four months then transferred to Motor Transport Repair Unit 315, where he served until discharged.[76]

Virgil Hertzog, a twenty-two-year-old drugstore clerk, also joined the army's gas hounds, enlisting at Clarksburg, Virginia, on 14 July 1918. He received training as an automobile mechanic at Middle Tennessee State Normal School in Murfreesboro, Tennessee, and later served with the Motor Transportation Corps (MTC), 310th Motor Repair Unit, in Coblenz, Germany. Hertzog recalled that his mechanical training was satisfactory, but unit discipline was rather lax: "We could care less—we were civilian 'soldiers'—to hell with the 'stripes'-'bars-'brass'-'silver'— whatever. . . . Ours [was] not a military outfit—strictly mechanical."[77]

The war also provided many thousands of soldiers outside the Motor Transportation Corps with their first training and experience with motor vehicles. The Signal Corps undertook the training of truck drivers and mechanics for its mobile communications units. The Ordnance Department, which maintained the army's tanks and other specialized vehicles, conducted a two-month motor course for officers and enlisted men in Raritan, New Jersey. The Medical Department sponsored a motor school at Camp Greenleaf, Georgia, and contracted with the Carnegie Institute of Technology for the training of medical officers and enlisted men in motor operation and maintenance for service in motorized ambulance units. The Coast Artillery School at Fort Monroe, Virginia, graduated more than ten thousand truck drivers during the course of the war. Letters home from Corp. Louis Chouinard reflect the novelty of this experience for many of those who served. His father owned a garage in Fall River, Massachusetts, but Louis evidently did not follow his father's trade, working instead as a carpenter before enlisting in December 1917. The army needed truck drivers, however, and Louis soon boasted to his father that he had become "some driver" and could now drive any car or truck and was learning all he could about repairing them.[78]

Hundreds of Americans outside of the formal army organization also gained new experience with motor vehicles during the war. The YMCA and the American Red Cross both relied heavily on trucks and automobiles to carry out their war welfare work. The YMCA used trucks to bring entertainment such as plays and movies to the soldiers at the various camps, both at home and overseas. In an effort not to burden the resources of the camps they visited, the YMCA trained many of its own personnel to handle the maintenance and minor repair of the trucks they used. The Red Cross used a large number of Ford cars and trucks to shuttle injured soldiers between medical facilities behind the lines. Many women learned how to drive, maintain, and repair their own vehicles while working for the Red Cross or for the various volunteer Women's Motor Corps established in American cities to help ease the domestic transportation crunch during the war.[79]

Demobilizing Wartime Mechanics

The postwar experiences of these Americans no doubt varied. David McNeal's diary ends with his discharge. He returned home to finish high school, but it is unclear if he used his new skills as an auto mechanic after graduation. A few other examples, however, provide at least a taste of how army motor skills translated into civilian life. James K. Moore, a farmer in Missouri before enlisting in the army, received training in the motor transport school at Washington University in St. Louis before serving with the 312th motor repair unit in Bordeaux, France. After his discharge Moore returned to farming, and when asked years later if he found his army training useful in civilian life, he replied that working on trucks for the army helped him to repair tractors and equipment on the farm.

Ernest Petrea, a farmer and public school teacher in Texas, volunteered before being drafted so that he could be trained in motor repair at Texas A&M College. Following his discharge, he did more farming and teaching and eventually opened a dealership to sell Chevrolet and Durant automobiles. Oscar Arneberg was already working as a tire man in a repair shop in North Dakota before enlisting and serving in the 309th Motor Repair Unit. Upon his return he was unable to get his old job back, so he opened his own successful tire and repair business.[80]

In contrast with these men's experiences, Lynn Snoddy, of Georgia, entered the army already possessing significant experience in automobile and truck repair. He received six weeks of training at the Alabama Polytechnic Institute in Auburn but claimed that he "already knew most of it." Consequently, although he went to work as an auto mechanic after discharge, he found his army experience to be of little use after the war.[81] William Rumbaugh, a twenty-one-year-old mechanic in Cleveland, enlisted in December 1918 in response to a "request by [an] officer who came to [the] factory to recruit certain qualified personnel." Rumbaugh underwent training at Fort Meigs, Washington, D.C., "only in necessary greenhorn instructions as regards marching, formations, etc." Years later he recalled his service during the war as being "a real detriment" when returning to civilian life.[82] The army did not generally offer professional development to experienced mechanics such as Lynn Snoddy or William Rumbaugh. Indeed, it would have been content solely to exploit the nation's population of experienced mechanics for the duration of the war, had there been enough such men. Yet the requirements of mechanized warfare and the novelty of automotive technology pushed the army to make mechanics out of farmers and clerks, blacks and whites, men and women.

When the war ended, the army disbanded the Committee on Education and Special Training and the network of classes it coordinated at the various public and private schools across the country. Military concerns about training did not cease, however, with the end of the war. Each of the twelve CEST district directors submitted final reports that were unanimous in their self-congratulations for a difficult job well done. The directors believed that they had made a significant contribution to vocational and trade education in America. James A. Pratt, district four educational director, believed that "the work at the various schools . . . [had] demonstrated the practicability of the short unit course to the educational profession of America."[83] R. W. Selvidge, district five educational director, reported his conviction "that with a going organization, well trained, experienced teachers, and adequate equipment you can train the man of average intelligence, between the ages of twenty-one and thirty, in less than six months until he will be at least the equal in skill of the average journeyman in that trade."[84] After filing their reports, these men returned to civilian life, carrying their experiences and convictions with them.[85]

Meanwhile, as the army began shipping tens of thousands of soldiers home, new training concerns emerged: recruiting enough soldier-mechanics to keep army trucks running during peacetime and retraining injured and disabled veterans returning from the war. To help recruit prospective mechanics, the Motor Transportation Corps consolidated its training operations at four camps: Camp Jessup near Atlanta, Camp Normoyle near San Antonio, Camp Boyd at El Paso, and Camp Holabird at Baltimore.[86] Then the army dangled automotive training before demobilizing vets to entice them to re-up for four more years. An MTC recruiting circular asked returning soldiers:

> Look around; se[e] what others have to offer; then consider what the Army, not the Army at War, but the Army at Peace, has to offer.
>
> They will pay you to learn a trade. The Motor Transportation Corps, the automobile arm of the Army, has the information you need . . .
>
> Never before has such a chance been offered to the youth of our country, and if you really want to attain the position of Expert Journey-man in the automotive industry, this is the chance you never had before.[87]

The MTC schools likely persuaded at least some vets to reenlist or new recruits to join. *Motor Age* claimed that Camps Jessup and Holabird had the largest and best-equipped automobile schools in the world and that their graduates were being "snapped-up" by employers.[88] Addressing the army's second concern, Congress charged the Federal Board for Vocational Education with coordinating the training of injured soldiers at private and public trade schools around the country—essentially a scaled-down version of the University of Uncle Sam system used during the war.[89]

The West Side YMCA Automobile School, like many other trade schools, catered to Federal Board students in the postwar years. Contemporary with the signing of the Armistice, enrollment at the school began to decline, and problems with the automobile curriculum became apparent.[90] The directors of the wartime CEST had expressed confidence in short-unit trade courses, but their frame of reference was the multiyear apprenticeship advocated by those in more traditional trades such as machinists and printers. The courses offered by the West Side auto school, comprising about ninety hours total class and shop time, were already too short even by army standards. The West Side education committee noted this discrepancy. It also eyed the large number of returning soldiers, some of whom who were showing up at the West Side branch "inquiring about courses of study and places for employment."[91] Thus, at a special meeting the West Side's education committee outlined a 732-hour Automobile Mechanics course encompassing the former auto shop, road work, and self-starter courses, a new Machine Operation and Shop Work course, plus an additional 546 hours of "practical repairing of automobiles."[92]

The expanded auto mechanics course breathed new life into the West Side auto school. Almost immediately, the Federal Board for Vocational Education began referring wounded soldiers and returning veterans to the West Side for retraining and transition to civilian life.[93] A few Federal Board students took business English or mechanical dentistry courses, but the vast majority came to the West Side for the complete two hundred–dollar automobile mechanics course.[94] The new course also attracted increased numbers of civilian students, and by 1920 the automobile and automobile repair courses accounted for more than two-thirds of the West Side's educational department revenues.[95]

The Knights of Columbus (K of C), a Roman Catholic fraternity that had provided food and entertainment to soldiers during the war, also began sponsoring automobile schools for returning soldiers after the war. With nineteen million dollars remaining in its war welfare fund at war's end, the Knights of Columbus opened its first evening school in Boston on 7 July 1919, offering trade and business classes to all former service men. Reluctance to attend Protestant YMCA schools motivated some former soldiers to attend the Knights classes. More important, the K of C charged no class fees to former servicemen, while civilians could and did enroll for a small fee. By the 1920–21 school year, the K of C conducted a system of 125 evening schools in thirty-one states, with a total enrollment of 99,310 students. The automobile mechanics courses proved the most popular of the 86 different courses offered, with a reported 55,000 men and women enrolled in 1920.[96]

No separate numbers survive on what portion of the K of C's 55,000 auto students were women or whether they took driving or mechanics' courses.[97] That the Knights offered free training to former servicemen likely placed women at a disadvantage where enrollments were tight. Nonetheless, women who wanted to continue or extend the mechanical experience they had begun as auxiliary drivers and mechanics during the war or who nurtured new interests because of the publicity given those women could have used the K of C, the YMCA, or other private auto schools to further their mechanical knowledge.[98] U.S. Census numbers from 1920 confirm that more than 1,200 women were interested enough to be counted as working in auto-related occupations in the transportation sector of the economy. The 1920 census did not yet report "automobile mechanic" as a distinct occupation; it did report, however, 949 women working as chauffeurs, 207 women working as "garage keepers and managers," and 111 female "garage laborers." These women represented less than one-half of 1 percent of the workforce in these occupations, but their numbers had increased twenty-fold since 1910, while men's relative numbers in those fields increased six-fold.[99]

The automobile courses remained the overwhelming choice of K of C students through the 1923–24 school year.[100] These high enrollments, as well as those at the YMCA and at a number of commercial and correspondence courses that continued to thrive after the war, indicate that enthusiasm for automotive knowledge and experience continued in the postwar years. During the 1920s this continued enthusiasm of young learners converged with employers' perceived need for more and better mechanics, with the growing momentum of Progressive Era educational reforms, and with newly available federal money to create vocational automotive programs in public schools. By the 1930s a new system for creating mechanics eventually eclipsed the private system of automotive education, with significant consequences for the auto mechanic's occupation.

CHAPTER 4

The Automobile in Public Education

In 1912 a group of boys at San Bernardino High School in California presented a petition to the board of education asking it to include classes in electrical work, gas engines, and automobile repair.[1] Four years later the high school in Duquesne, Pennsylvania, organized a class in vulcanizing automobile tires "in response to a request from the boys themselves."[2] When the high school in Sioux City, Iowa, opened an auto mechanics class in the fall of 1920, school officials found that "there were more boys on hand than we could accommodate." The next year they devoted more space and equipment to the auto shop and, symbolic of the new age, evicted the wood shop from its lathe room and turned the space over to the auto shop.[3] In Riverside, California, students in the vocational auto shop class at Polytechnic High School channeled their enthusiasm about their class and subject by organizing an extracurricular "Mechanics Club" in November 1923 for the purpose of creating "a closer relationship among the students and between the students and instructor."[4] Such enthusiasm among students for automobiles was so commonly recognized in the 1920s that many educators saw auto shop as a way to "keep the boys interested in school."[5] As one advocate of auto shop wrote, "All red blooded American boys are interested in automobiles."[6]

During the 1920s automotive education in public schools supplanted the system of private education typified by the West Side YMCA's auto school. As the West Side auto school approached its twentieth anniversary, enrollments once again declined, and in early 1923 the education committee referred to the auto school as "our greatest concern."[7] Weak enrollments forced the directors to cut expenses, fire instructors, and consolidate operations into one building.[8] The education committee's minutes of May 1921 noted, "The Educational Work of the Association [is] constantly in competition with public and endowed schools where the tuition can be either free or greatly under that of the Association."[9] The members consoled themselves with the fact that other YMCA auto schools also faced difficult times and that nineteen chapters had closed their auto schools during the

1921–22 season. They were not alone.[10] As young students gained the knowledge they sought at no cost, or more accurately, as school boards decided to use tax dollars to create mechanics, public vocational education eclipsed the early private and nonprofit auto schools as the primary source for creating new mechanics.

Thus, a new and relatively stable system for creating mechanics grew in conjunction with vocational education programs in America's public schools.[11] Public vocational education prepared thousands of boys for eventual entry into the trade through a system that endured essentially unchanged for most of the remainder of the twentieth century. This is not to say that all mechanics, nor even necessarily good mechanics, came out of the public school system. All of the major automakers maintained their own in-house programs for training dealer mechanics that were much more intensive and specific than vocational education courses. Rather, automobile vocational education, or "auto shop," helped boys decide whether to become mechanics and aided those who chose to do so in forming their social identity as mechanics.

Furthermore, by displacing private auto courses, auto shop cut off alternative avenues of access to automotive technological knowledge. It further narrowed and ossified the already strong gender and class construction of the trade. While tax-supported auto shop provided more African Americans with access to technological knowledge, it denied equal access to those in segregated school systems. At the same time, the institutional apparatus of vocational education effectively barred girls from auto shop, and thus from the trade, and made auto repair the preserve of working-class boys who were "good with their hands," rather than "college-bound." Automobile vocational education in the United States sorted adolescents and fed a significant number of boys into the auto mechanics trade with some technical training but, more important, with a major part of their socialization accomplished. They could then be further trained in the technical details of the trade either on the job or at factory seminars. In short, high school auto shop reflected and often sharpened broader social hierarchies and intertwined them with a particular knowledge of and relationship to automotive technology.

Four factors converged in the 1920s to create the long-lasting institution known as high school auto shop. First, the concept of vocational education began to take hold in public schools. Educational reformers had been working for more than two decades, with mixed success, to make public schools more meaningful and attractive to non-college-bound students and their parents. By introducing manual training and eventually vocational education to the curriculum, reformers hoped to reach disaffected students, keep them in school longer, and prepare them better for their life's work. Second, in response to these Progressive Era educational reforms, Congress passed and President Woodrow Wilson signed the Smith-Hughes Act in 1917, providing federal matching funds to the states for the express purpose of promoting vocational, agricultural, and home economics education. Third, automobile manufacturers and their dealers who employed large numbers of mechanics turned to the topic of training more and better mechanics as one means for dealing with the "problem of service" during the 1920s. Finally, as we have seen, students themselves maintained and expressed intense interest in automobile technology, at times urging local school boards to establish automobile courses. Together these forces established auto shop courses in public schools around the country during the 1920s and into the 1930s.

Educational Reform in the Progressive Era

The conflicting goals of educational reformers in the years prior to mass automobility set the stage for high school auto shop to become a sorting and training ground for future auto mechanics. The origin of these conflicting goals reaches back to the late-nineteenth-century debate over "manual training." As an educational reform, manual training advocated using the hand to educate the mind. To reformers such as John Dewey, John D. Runkle and Calvin M. Woodward, manual training provided a way to liberalize teaching methods in America's schools and to get away from rote memorization and the recitation of "lessons." Instead, students would be exposed to the products and processes of modern life, and through students' "natural" interest in these "real world" items, teachers would guide pupils to higher levels of learning. Manual training would be beneficial for all children, rich or poor, as the intended outcome was higher-level learning and understanding of the "real world."[12]

Others, however, saw manual training primarily as a means to teach nontraditional students basic academic skills together with useful industrial skills. For much of the nineteenth century public education was not widely available, and high schools served that very small percentage of youth whose families could afford to forgo their labor or wages while they prepared for college. Skilled apprenticeships declined dramatically over the nineteenth century, while increasing numbers of boys and girls worked at unskilled jobs at an early age.

As Americans reevaluated the role of children and adolescents in society during the late nineteenth and early twentieth centuries, a consensus began to emerge that youths needed better protections and guidance to survive and thrive in industrial society. The public school seemed to be the ideal place to accomplish this. Thus, along with the passage of better-known anti-child-labor laws, many states passed compulsory school attendance laws. Not only did progressive reformers want to restrict the type of industry and age at which young people could work; they also wanted to get them into the controlled environment of the public school.

The resulting influx of nontraditional students changed the dynamics of public education in America and complicated the mission of public high schools. In Chicago, for example, average daily attendance in the public schools jumped from 25,300 students in 1870 to 231,400 in 1900.[13] Manual training, some said, would serve to interest these new, nontraditional students and keep them in school, where they could learn important lessons of citizenship and culture. By teaching a boy to use a hammer and anvil, one might also be able to slip in a reading of Longfellow's poem, "The Village Blacksmith," with its celebration of diligent toil and perseverance. In this manner immigrant children could be "Americanized," and working-class children would learn middle-class values such as thrift, punctuality, and respect for authority.[14]

These two approaches to manual training—as a means to liberalize curriculum for all students or as a separate curriculum for nontraditional students—reflected the tensions between the ideal of a truly democratic public education and the pragmatic belief that not all children are destined for academic achievement. Was manual training to be for everyone or just for the poor? Despite its growing presence in schools after 1880, critics attacked manual training as, on the one hand, economically irrelevant if it did not prepare students for specific employment or, on the other hand, undemocratic if it looked too "vocational." Calls for "practical" industrial and vocational education soon drowned out the latter criticism.[15]

The manual training movement smoothed the transition to vocational education in public schools by changing the conception of what might legitimately be taught. Americans increasingly accepted the educational goals of vocational preparation and industrial job skills as they embraced the social and economic changes associated with turn-of-the-century industrialization. In their 1974 study, *American Education and Vocationalism,* Marvin Lazerson and W. Norton Grubb note that "between 1890 and 1910, vocational education attracted the support of almost every group in the country with an interest in education."[16] Many educators embraced vocational education for the same reason that some had advocated manual training: as a means for reaching and serving the steadily increasing numbers of nontraditional students in public high schools.

Industrial employers, through the National Association of Manufacturers (NAM), strongly supported the move toward vocational education. Public vocational education would fill the void of skilled labor left by the demise of apprenticeships. More important, vocational education would help semiskilled and factory workers understand their role in the production process, in the economy, and in society and thereby stave off the urge to join radical labor unions. The NAM advocated establishing public vocational and technical high schools supported by tax dollars but separately administered by "practical men of business" rather than "education men." Such a separate system of control, they believed, would ensure that vocational curriculum would not drift toward irrelevance as they believed manual training had previously done.[17]

Organized labor initially resisted trade training and vocational education; sensing, however, the strong trend toward public vocational education, the American Federation of Labor finally embraced the movement. Many union locals ran their own apprentice training programs through which they could control entry into a given trade. With some justification they viewed employer-controlled trade and vocational schools as "scab hatcheries." They adamantly opposed the separate administration plan advocated by the NAM and feared that even in the hands of professional educators vocational education ran the risk of creating a stratified school system that would hamper worker mobility in the end. Unable to recapture the guild-controlled training ideal of centuries past, important segments of organized labor embraced public vocational education and worked to shape its implementation to the worker's favor. This seemed to them the best defense of democratic workplace values.[18]

With educators, industry, and organized labor agreeing on some form of publicly supported vocational education, advocates organized the National Society for the Promotion of Industrial Education in 1906 and in 1917 secured passage of the Smith-Hughes Act, which provided federal matching funds to the states for establishing vocational education programs in public school systems. The legislation sidestepped thorny issues of exactly how to implement or administer such programs, requiring only that the money be used to support agricultural education, trade and industrial education, and home economics education.[19] The act set aside just over $1.6 million in its first year, with annual increases to just over $7 million in 1927, and established the Federal Board for Vocational Education to oversee the distribution of these funds. States and local school districts wishing to share in this unprecedented federal interest in vocational education had to make dollar-for-dollar matching investments in their public vocational education programs.[20] Such programs could be full-time day schools, part-time schools, or evening "continuation" schools for workers who wanted to advance in their trade. By the time the United States entered World War I, much of the groundwork for establishing vocational education in the public schools had been laid. It had a broad base of support and, for the first time, federal funding to spur more state and local school systems to develop their own vocational education programs.

While World War I brought national attention to the need for more automobile and truck mechanics, it also gave an urgency and focus to the broader vocational education movement. On 9 November 1917 the Federal Board for Vocational Education authorized payment of Smith-Hughes money to public school systems instituting mechanical and technical training for conscripted men.[21] Using this opportunity, some local school districts jump-started or expanded their vocational education programs during the war years to help train soldiers in various trades as part of the University of Uncle Sam system. In Los Angeles, for example, five high schools trained 675 soldiers in twenty-nine different trades during the war. The Newark, New Jersey, board of education organized wartime classes at their Central High School "for auto-mechanics, bench wood-workers, blacksmiths, carpenters, electricians, machinists and sheetmetal workers."[22] In Chicago the board of education secured "very complete" equipment for training soldiers at the South Division School "with the intention of retaining a permanent auto mechanics school."[23] Automobile dealers and garage owners in Cincinnati worked together with the board of education to open a Smith-Hughes funded school to train soldier-mechanics in 1918. Following the war, the board purchased the school equipment outright and opened the Automotive Trades School of Cincinnati, and by late 1920 six other Ohio cities were reported to be opening automotive trade schools.[24]

In all, twenty-six public high schools appear in the final reports of the Army's Committee on Education and Special Training (CEST) as having formally participated in the University of Uncle Sam war-training efforts, many of them training auto mechanics. The CEST preferred to contract with schools that had nearby housing for the soldiers, so colleges, universities, and private schools conducted much of the soldier training. It is likely that a great many more public high schools, encouraged by the Federal Board for Vocational Education, organized soldier training programs but were not ultimately selected to participate officially in the training of conscripted men.[25]

Putting the Auto in Auto Shop

After the war automakers and dealers anticipated the resumption of strong prewar new car sales, only to find the economy in a sharp downturn following rapid war demobilization. New car sales stagnated in 1920–21, and automakers and dealers began to consider the prospect of a saturated new car market in which replacement sales would be as important as first-time purchases. In such a market service would play an important role in convincing buyers to stick with the same car make or to switch to another, more reliable maker or dealer.

Consequently, in the 1920s many automakers and industry observers began to speak of "the problem of service." They recognized that from the motorist's perspective, using the technology on a daily basis meant dealing with mechanics and garages for regular maintenance and occasional (perhaps frequent) repairs. Those encounters with mechanics and repair shop personnel were often fraught with anxiety. For many of the same reasons that wealthy urban motorists had previously found it difficult to supervise their chauffeur-mechanics, motorists in the postwar years continued to fear that incompetent or dishonest mechanics and garages would run up unnecessarily high repair bills. As one writer put it at the time: "It is no wonder that the motorist who has driven for a couple of years acquires little lines around the eyes and mouth. It is not all brisk breezes driving against the wind which bring the lines. Some of them are from keeping up with Ali Baba and his Forty Thieves."[26] Automobile technology had become more reliable than in the chauffeur-mechanic days; it had also become more sophisticated, however, and many motorists remained incapable of either diagnosing the symptoms of their own automobiles or accurately judging the veracity of a mechanic's diagnosis and treatment.

During the 1920s automakers and dealers increased their efforts to train competent and reliable mechanics as one way to address the service problem. In Minnesota the Garage Owners' Association worked out a training and placement agreement with the Dunwoody Institute with the goal of producing "more and better mechanics," reducing costs, and improving automotive service in that city.[27] On the national level the National Automobile Dealers Association began an educational campaign in cooperation with the manufacturers of component parts such as Continental Motors, the Timken-Detroit Axle Company, and the Borg and Beck Clutch Company. It consisted of a series of traveling "service talks" to train mechanics in various cities in the proper adjustment and repair of key mechanical units.[28] The Ohio-Buick Company of Cleveland went so far as to establish a thirty-three-month apprenticeship program for seventeen- to twenty-five-year-old boys. In the spirit of traditional apprenticeships the company claimed it monitored and influenced the "companions and habits" of its young charges. Betraying objectives beyond merely technical competence and echoing the earlier language of the YMCA's auto school, J. F. McDonald, general service manager of Ohio-Buick, said of his program, "Education causes us to be courteous and polite, to use proper language in the proper place, to be careful about our clothes and our physical condition; in fact, it makes us better men in every way."[29] Not all employer-run training programs had such high social aims, yet all involved considerable expense on the part of the employer and with no guarantee that graduates would remain with the companies that trained them.

Some in the industry soon realized that they had much in common with advocates of vocational education. In much the same way that wealthy motorists turned to the New York YMCA for a "better class" of chauffeurs, automakers and dealers looked to the public schools to produce more mechanics with proper work habits and attitudes. In the schools, thanks to educational reformers who had preceded them, they found tax dollars earmarked for vocational education, examples of schools that had already adopted automobile courses, and an increasing literature on automotive course objectives, outlines, and methods.[30] All that remained was to convince more local school districts to include automobile courses in their vocational programs. *Motor Age* urged dealers in 1920 to "Get behind Your Public Schools." "One of the most gratifying things coming to light these days," wrote the editor, "is the tendency on the part of public schools throughout the country to add to their curriculums, a course in automotive mechanics. [This development] is something dealers should get squarely behind and lend their full support."[31]

John Calvin Wright, director of the Federal Board for Vocational Education, former educational director of the Rahe Auto and Tractor School in Kansas City, Missouri, and author of a popular automotive repair textbook, urged attendees at the 1924 National Automotive Service Convention in Detroit to cooperate with their public schools. "The public schools cost $2,000,000,000 a year. They must be looked to more and more for assistance in the general solution of the automotive maintenance problem."[32] Roland Cummins, a manager at Packard's Philadelphia branch, did just that. He believed that the cost to the industry of training the men needed would add "thousands upon thousands of dollars" to dealers' overhead. If, however, men in the trade and boys entering the trade could be trained in the public schools, "such action would be greatly appreciated and very enthusiastically greeted by the automobile businessmen." Accordingly, Cummins worked with the Motor Truck Association of Philadelphia, the Philadelphia Automobile Trade Association, and the head of vocational education in Philadelphia schools to introduce a course in automobile and truck mechanics in that city's high schools.[33] In Cleveland the members of the Automotive Association of the Chamber of Commerce asked that city's board of education to seek federal and state funding under the Smith-Hughes law "to offer instruction to school pupils in automotive mechanics." With such curriculum in the public schools, they believed, "the automobile mechanic will be given new dignity, more young men will turn to these trades and in the future the industry will have more skilled mechanics to select from."[34]

Enthusiasm for bringing the automobile into the public school was near universal, but implementation of any new educational curriculum or reform is a notoriously local and particular affair; auto shop was no different.[35] Each community felt different needs and had differing resources to draw upon when considering adding automobile courses to their public schools. In general, however, the courses offered in public schools across the country from the 1920s down to the present can be divided into four types: prevocational courses, one-semester "home mechanics" classes, evening continuation schools, and full- or part-time vocational trade courses.

Prevocational automobile classes, offered at the junior high or intermediate level school, resembled most closely the old manual training ideal. Young boys would rotate among various "industrial arts" shops during the year in order to gain "an appreciative understanding of the problems involved in the work of the world." These "exposure experiences" would help boys make "earlier and wiser choices of vocations and training courses," thereby reducing the number of "misfits" in the workplace. Such prevocational automobile courses did not seek to produce employable mechanics but merely to teach the general parts and operation of an automobile and to convey a sense of the work of a mechanic.[36]

Single-semester courses in regular high schools aimed at teaching boys how to care for their fathers' cars and eventually their own.[37] This type of automobile course was often part of the "home mechanics" movement that sought to teach boys a smattering of skills thought to be useful at home, a masculine counterpart to home economics. The single-semester course aimed not at training boys to become auto mechanics but, rather, at creating "intelligent owners and drivers" and, presumably, better customers of mechanical services.[38] Such a course "is especially valuable," wrote one manual arts educator, "because it teaches about a subject which will be associated in the life of almost every man."[39] Work in these short courses, like in the prevocational courses, focused on the operation, care, and proper methods of repair of automobiles but did not teach "skill and speed." Similar in many ways to the automobile owners' courses offered by the YMCA, some high schools also offered this type of course at night for adults in the community.[40]

Smith-Hughes money encouraged schools to establish another type of automobile course at night, the evening continuation school. States could seek federal matching funds for evening programs to help workers advance in their trades. Therefore, some urban schools maximized the use of their auto shop equipment by offering short, specific trade extension classes to employees of local garages and dealers. Men enrolled in these courses to gain specific knowledge they thought would help them in their work. In response organizers tried to offer short "unit courses" in ignition or brakes or lighting and electrical systems or some other discrete subject rather than general theory or methods of repair. In order to receive federal aid for these courses, schools had to restrict enrollment to men already in the trade, so evening continuation courses generally did not create many new mechanics.

More significant to the creation of new automobile mechanics were the vocational trade courses of one to four years in length, and the cooperative programs in which boys worked in a garage or shop part-time and attended school part-time. Organizers of these courses sought to produce more and better mechanics, just as their future employers desired. Often taught at vocational high schools or technical high schools, separated physically and symbolically from the regular high school, educators aimed these courses at "mechanically inclined" boys intending to go into the auto mechanics' trade. An examination of available curricula for these vocational auto shop courses reveals teachers' and administrators' expectations about their students' relationship with technology and about their future role in the workplace and society. Vocational auto shop courses generally prepared students to enter a departmentalized workplace in which the pace of work and even the order of operations might be determined by forces beyond their control.

Virtually all auto shop curricula broke the complex technology of automobiles down into functional systems, each to be studied and mastered separately. California's State Board of Education, for example, suggested a scheme of sections and units that taught brakes, suspension, and springs separately from clutch, transmission, and rear axle as "best for training a learner who desires a complete knowledge of the trade." At the Automotive Trades School in Cincinnati, principal Ray F. Kuns shifted first-year students "from one department to another" every ten weeks. Students rotated among the chassis department, engine department, electrical department, battery department, and tire-vulcanizing department before being assigned to the service department, where they would work on cars belonging to the board of education and the general public.[41]

This division of a complex technology into subsystems provided not only a sensible pedagogical method; it also helped prepare students for the divided, rationalized workplace they might find if they worked for a large, urban repair shop. Beginning in the late 1910s and continuing through the 1920s, many large urban garages divided their shops into engine-rebuilding departments, brake departments, battery departments, and so forth, employing mechanics with specific abilities or experience for a given department.[42] Both Ford and Chevrolet urged their dealers to adopt departmentalization and mechanic specialization in their shops.[43] A full-page newspaper advertisement for a large Washington, D.C., garage named Call Carl's in the late 1920s reveals the extensive degree of departmentalization in some shops. Bordering the page are photos of eleven departments, from a blacksmith department to an upholstery department, in one multistory building.[44] This type of departmentalization aimed at reducing labor costs and making service more "efficient," and it would have been familiar to students coming out of most auto shop courses.

Vocational auto shop also acclimated students to the demands that systematic management and the "flat rate" system made upon workers. During the late 1910s and much of the 1920s industry sages promised that systematic management methods—including accounting, inventory tracking, and job costing—would staunch the flow of red ink that many dealers experienced in their service departments.[45] Instituting bureaucratic controls and reforms in the shop would help put service "on a paying basis." Smaller, independent shops would also benefit from tighter management procedures, as many proprietors were good mechanics but not good businessmen. We see this concern with accounting and "system in the shop" reflected in the requisition forms, "job sheets," and time sheets used by some shops and touted by the automotive trade press.[46] In vocational auto shop classes instructors typically expected students to fill out and submit similar forms in the course of their shop work. They apparently did not teach their students what to do with that information or how to analyze it for managerial purposes. They simply trained them to fill out the forms and pass the information up the chain of command.[47]

The common use of time records and instruction sheets in school shop classes in general indicates the degree to which many shop teachers took their cues from industry practices.[48] The teacher of a nonautomotive shop class in Wisconsin wrote specifically of adopting "the Taylor system of management" in his school shop. If a boy did not finish an assigned job within the time set, "the piece was thrown away and a fresh start made." Furthermore, the student had to perform the operations "in the order they were given in the instructions" and "in the manner called for in the instructions." In this way the instructor believed he could educate "a greater number of satisfactory employes [*sic*] as well as better workmen."[49] It is unclear if many auto shop instructors provided students with flat rate times or target times for their shop work, but students would certainly have learned that time mattered in the shop.[50]

Auto shop instructors commonly required students to fill out time sheets on their jobs. Some instructors used that information for both assessment of the student's progress and for billing of work done on cars from the community. In the Automotive Department of the Boys' Technical High School in Milwaukee, job times were recorded for billing purposes on the back of a "Repair Tag" attached to all cars brought into the shop for service. In addition, the instructor kept a record of jobs performed by each student in the class, the time the student spent completing each job, and his grade on that task. The "Auto Shop Pupil Record" recorded the student's cumulative grades for each marking period with overall ratings for the "Quality" and "Speed" of their "Workmanship."[51] Auto shop instructors also relied on job sheets that gave step-by-step instructions on how to do a certain job.[52] Such preprinted instructions would have prepared students to accept the idea that there was indeed "one best way" to do each job, a major premise of the flat rate system and of scientific management.

Vocational auto shop also prepared students to become auto mechanics by what it did not teach them. The public school system did not prepare auto shop graduates to create or design automobiles, only to fix them. Rarely did vocational auto shop curriculum include more than scant math, science, or other academic subjects. Most required only as much academic preparation as would be directly related to a mechanic's job. The Los Angeles City School District listed "automotive engineer" as one of the possible occupational outcomes of its auto shop programs in 1921.[53] The California State Board of Education asserted, however, that little of the academic work beyond the eighth grade applied directly to the auto mechanic's trade. It considered such high school staples as science, mathematics, and English to be "supplemental subjects" for vocational students, "not as subjects to be studied in and for themselves." Supplemental science for auto shop students, for example, "should not include the systematic study of any of the sciences as a whole, nor should it conform to the requirements of a general science course. The science problems of the occupation constitute the basic material for this work." Similarly, prospective auto mechanics needed only limited mathematics education: "The problems of science mentioned above will in many instances have to be solved by derived formulas [such as rating the horsepower of an engine]. The derivation of such a formula should not be made a subject of study. . . . Process work in algebra should not go beyond fractional equations. Only such facts of geometry as are necessary for the solution of the mechanical problems of the trade . . . should be taught."[54] The Boys' Technical High School in Milwaukee required automotive students to take "special allied courses" in English, mathematics, science, and mechanical drawing. The purpose of the "automotive trade mathematics" course was "to acquaint [the automotive student] with the simpler arithmetical processes that he must be able to handle in order to be successful in his trade." By contrast, architectural drafting students at the same Milwaukee school took allied courses in English, arithmetic, algebra, trigonometry and logarithms, chemistry, physics, slide rule, and strength of materials.[55] The absence of high school–level mathematics, science, and English in most auto shop programs meant that graduates would not be qualified to enter college or pursue a professional career without significant remedial work (see fig. 14).

Furthermore, vocational auto shop courses did not typically teach aspiring mechanics business skills, even though most mechanics in the trade at the time likely worked for small shops, and many seem to have harbored desires to open their own shops.[56] Vocational auto shop courses in the public schools rarely, if ever, included supplemental course work in bookkeeping, accounting, or other business skills that might have helped students who wished to strike out on their own as young entrepreneurs.[57] Several explanations might be offered for this absence. First, the students themselves may not have been interested in business skills classes, though this alone would hardly have been ample justification for omitting the subject. Another explanation is that educators knew, even after a two-year vocational auto shop course, that they were not graduating knowledgeable or experienced mechanics. Their young graduates would still need years of on-the-job experience before they could be considered "journeymen" auto mechanics. Therefore, why teach them business skills? They would not be competent to open their own shop for many more years, and even then only a fraction would choose to do so. Finally, local garage men and dealers who looked to the high school auto shop for entry-level employees would not have looked favorably on a program that prepared or encouraged graduates to establish competing auto repair shops.[58] A decade earlier private automobile schools may have touted the potential independence of the auto mechanic's trade in order to attract paying students, but in the 1920s and 1930s preparing students for independent proprietorship was not the aim of auto shop in most public school systems.

Guiding Boys into Auto Shop

Youthful enthusiasm for automotive technology ensured that educators did not have to work hard to fill auto shop classes during the 1920s and 1930s. Yet educators did have specific expectations about what types of students would, and should, take vocational auto shop. Echoing one of the rationales behind manual training, educators believed these courses would appeal to and were appropriate for "over age boys," dropouts, and boys who would have to leave high school to go to work.[59] These expectations, formalized by the new "science" of vocational guidance, helped make high school auto shop the almost exclusive preserve of working-class boys.

Vocational guidance had its origins in the settlement houses of the Progressive Era. Young job seekers sought assistance from settlement houses such as the Henry Street Settlement in New York City and Hull House in Chicago. Soon settlement house workers "found themselves deeply involved in the hopes and trials of job-getting."[60] In response, some settlement houses established vocational bureaus in an attempt to coordinate job placement and job training for working children. The Civic Service Houses of Boston established the Boston Vocational Bureau in 1901 and put Frank Parsons in charge. Parsons, known to some as the father of vocational guidance, used personal interviews and various tests to determine a job seeker's qualifications and aptitudes. He then guided them to employment that he believed suited their profile. Such guidance, however, was not unbiased, as education historian Paul Violas recounts:

> An interesting example of the kind of profile developed by Parsons was "the case of the would-be doctor," described in his *Choosing a Vocation*. Parsons persuaded the client that medicine was an undesirable goal because of his humble social and economic background and also because of his race. These criteria weighed heavily against working-class and immigrant clients because the Anglo-Saxon middle-class model defined Parsons's ideal for higher status and better-paying vocations. Working-class children were thus discouraged from aspiring to prestigious callings and made to believe that they had been "scientifically" selected for low-level occupations.[61]

The bias exhibited by Parsons—no doubt born of compassion but shaped by historical context—permeated the vocational guidance movement as it shifted focus in the 1910s from job placement to educational guidance and moved from settlement houses into the public schools. In the schools guidance counselors and educators, embodying many of the same Anglo-Saxon middle-class values, sought to reduce the number of "misfits" in industrial society by guiding students to occupations for which their aptitude and abilities suited them. According to H. E. Stone, boys' counselor at Central High School in Erie, Pennsylvania: "Hundreds of cities, including Erie, have at last recognized the responsibility of the schools in preventing at least some failures in life, thru occupational and educational counsel that will reduce the number of misfits. . . . [E]ach student has his own distinctive qualities and abilities. It is fortunate, therefore, that each student in the high schools will have special help in adjusting himself to the right course of study and to the right life aims."[62]

Counselors helped students learn those "right life aims" using a variety of methods ranging from student interest inventories to mental testing and conveyed their findings in sometimes forceful, sometimes subtle ways. Most writers on the subject advocated using a battery of mental, aptitude, and interest tests to evaluate students and help them choose a vocation. Some advocated a strong, assertive role for the counselor, believing students needed to be told what they were suited for based on their mental, social, and physical abilities as well as their family background and the needs of the industrial economy. Professor Ray Simpson of Eastern Illinois Teachers' College exhorted guidance counselors and educators, "Give the boy the benefit of your judgment. . . . We allow too much drifting. The boy should be led into the field for which his intelligence, environmental condition, and special aptitudes have fitted him."[63]

Other writers took a less assertive approach to vocational guidance, instead advising counselors and educators to let students develop their own interests. Industrial arts programs in the junior high school or "occupation classes" in grade school would allow students to experience a variety of tools, materials, and subjects, thus helping them rule out obvious "misfits" for themselves. Based on these experiences, as well as their academic experiences and the "objective" career information provided by the school, children and their parents could decide which educational and career path to follow. Meyer Bloomfield, an influential voice in the early vocational guidance movement, wrote: "It is not the business of vocational advising to favor or disfavor occupations. It is primarily its function to know the facts, analyze, classify, simplify, and apply them, where they will do the most good. Let the facts speak for themselves, but drive them home. The responsibility rests on the shoulders of those who make the decision. But knowing the facts is no child's play, neither is their skillful dissemination."[64] Violas notes that such "skillful dissemination" of facts and self-selection could be even more effective than the Simpson-style assertive approach: "Someone coerced into an unsuitable job might blame his plight on the school, the state, or the economic order. If, however, a person could be led to believe that the decision 'came from within,' he would more likely accept responsibility for his condition."[65] It is likely that many schools and counselors used aspects of both approaches, similar to the North Bennet Street Industrial School of Boston, where educators combined vocational-information classes and individual student counseling.[66]

Some vocational guidance advocates placed more emphasis on intelligence testing as a means for counseling vocational students. "When the boy has selected his vocation," wrote E. Joseph Goulart, "we can direct his attention to branches in which his mentality would ensure success by using his intelligence quotient and the 'intelligence standards for occupations of 36,500 soldiers' as the basis of our suggestions."[67] Goulart referred to the Occupation Intelligence Standards developed by the U.S. Army's study of World War I soldiers which found certain ranges of intelligence correlated with certain occupations. Not surprisingly, most of the mechanical trades corresponded to average intelligence, with "general auto repairmen" ranking significantly below engineers, accountants, and YMCA secretaries but above laborers, tailors, and teamsters. As is the case with intelligence tests today, some critics of the time questioned the validity of mental testing, but in cases in which student IQ scores were available, it would have been difficult for counselors and educators to ignore them entirely when counseling a child. Some believed that such scores should be used at least to caution students whose occupational interests much exceeded their "intelligence" or to prod those of higher intelligence who set their ambitions too low.[68]

When it came to auto shop students, educators generally held low academic expectations. Lewis Wood expressed well the attitude that many of his fellow educators held toward vocational auto shop courses. Citing government reports showing that many students who entered high school did not graduate and that of those who did few went on to college, Wood wrote: "This means that over 90 per cent [of our children] must face the world with not more than a high school education. For these our shops and particularly our auto shops have a very distinct vocational value. . . . The boy who finds that he must go to work and has developed an interest in mechanics may spend more of his time in the shop and specialize along some line of work. For him the auto shop may be a pretrade course." Furthermore, according to Wood, "over-age boys not yet in the high school" (i.e., those who did not complete grade school) should be allowed to enroll in the high school auto shop: "Many of these boys are mechanically inclined and are apt to go to work in garages or follow other mechanical pursuits. . . . In our school shops these over-age boys, from whom many of our future mechanics come, will make a systematic study of the automobile which should give them a different attitude toward car work."[69]

Conversely, educator Harry Anderson expressed concern about letting too bright of a student pursue the auto mechanic's trade. "There will be found in a class," wrote Anderson, "many aggressive young fellows who would aim higher and whose ambition could be aroused to higher things, if the opportunities were pictured to them." An auto shop course "should portray the large field of opportunity for gainful employment in the higher class of work." Ironically, the higher class of work which Anderson had in mind was car salesman, an occupation with its own image problems relating to trust. For Anderson, however, selling cars was self-evidently superior to fixing them, and auto shop instructors should prod "aggressive young fellows" to this higher class of work.[70]

An anonymous photographer for the U.S. Office of Education in 1940 captured vividly the increasingly evident working-class association of auto shop courses (see figs. 15 and 16). The hats, overalls, and even physique of the auto shop students contrasts sharply with the two electrical shop students. A subtle hierarchy had emerged in public education, mirroring Western society's long tradition of associating abstract "head work" with social privilege and tactile manual work with the lower classes. Electricians used "instruments"—voltmeters and ohmmeters—which translated electrical properties into abstract numerical representations. Those numbers could then be placed into mathematical equations for diagnosing, predicting, and manipulating the behavior of the circuit. This abstract, representational knowledge was fundamental to understanding electricity-based technologies such as radio and formed the material basis for the social divergence of auto shop and electrical shop. In short, electrical shop students had to know how to manipulate equations; auto shop students did not. By comparison, the curricula for electricity-based technology courses generally included more science and considerably more mathematical work than auto shop programs.[71]

In this context of truncated curricula, more or less coercive counseling, and educators' low expectations of auto shop students, it is not surprising that auto shop became the preserve of working-class boys. Vocational education in general was for students who had shown "sufficiently marked tendencies to cause their separation from the regular classes,"[72] and auto shop in particular had one of the least demanding academic programs. The question of what kind of students actually did enroll in the various auto shop courses offered in public schools during the 1920s and 1930s warrants further investigation if sufficient sources of evidence become available. It seems likely, however, that educators got what they expected in auto shop students. Ray Kuns, principal of Cincinnati's Automotive Trades School, reported in 1923 that out of eight classes of students enrolled in his school's two-year day course (approximately two hundred boys total), "one class is usually of eighth grade caliber, six are of ninth grade caliber, and the other class is made up of boys of tenth, eleventh, and twelfth year ability."[73] That seven-eighths of Kuns's auto shop students were of "ninth grade caliber" or below is significant given that the Cincinnati school was one of the most respected and publicized public automotive schools of the period and could likely have drawn better students than most other auto shop programs.

It seems likely that well-intentioned educators presented with curricula stratified between specific vocational programs as well as between vocational and regular high school programs would have been inclined to guide working-class boys who were "good with their hands" into vocational auto shop, in which academic skills would be less important. On the other hand, middle- and upper-class boys with mechanical inclinations, if not discouraged from vocational courses altogether, would be directed to programs such as electrical shop or architectural drafting which might prepare them better for a "higher class of work," perhaps even college.

Segregated Auto Shops and Divided Highways

Young white males did not hold a monopoly on interest in automotive technology in public high schools. Besides the cumulative effect that curriculum and guidance decisions likely had on the social class and academic abilities of auto shop students, the institution of high school auto shop had the additional effect of increasing the gender segregation of automotive knowledge and supporting the racial segregation of American highways. As early as 1921, the Girl Scouts of America began offering the badge of the Winged Wheel to eighteen-year-old scouts who already held the first aid badge and could demonstrate proficiency behind the wheel. Modeled after the experiences of women's motor corps and Red Cross ambulance drivers during World War I, the new badge symbolized the vibrant interest many girls and women continued to have in automotive knowledge.[74] In spite of this, girls would have found it increasingly difficult to pursue their interest in motor vehicles as high school auto shop rapidly eclipsed other venues for obtaining automotive mechanical knowledge.

Girls of any social class were very unlikely to enroll in auto shop. Of the 5,926 students in thirty-seven states reported as enrolled in federally supported public school auto mechanics courses in 1920, only 6 were female. Looking further, of the 44,598 students identified as enrolled in Smith-Hughes-supported auto mechanics programs in three sample years (1920, 1930, and 1937), only 42 (less than one-tenth of 1 percent) were female, and none were enrolled in all-day vocational programs or part-time trade extension courses.[75] Girls' underrepresentation in high school auto shop may have been due, in part, to the exaggerated peer pressure that adolescents felt to conform to gender norms. Yet even when their interests overcame that pressure, school administrators likely denied girls the opportunity to enroll in auto shop. Shop teacher Lewis Wood admitted that "girls can profit" from a general auto shop course as they, too, would be future motorists: "However, very, very few will ever do more than change a tire, and it would hardly seem necessary to spend even a semester in the school shop when that is about all the use that will be made of what is learned. Then, too, if the girls are really to do this work, they must wear coveralls and get into dirt and grease, and the ordinary shop is not arranged with the necessary dressing and washing accommodations for girls."[76] Thus, by Wood's account not just social but institutional and even architectural barriers kept girls out of high school auto shops. Previously, at least some women had enrolled in private auto schools such as those offered by the YMCA and the Knights of Columbus. Rosalie Jones, the New York suffragist-mechanic whose example was supposed to shame blacksmiths into taking up auto work in 1915, would have found it very difficult, however, to gain access to high school auto shop. As public vocational education supplanted private auto schools over the course of the 1920s and 1930s, sources for women to gain automotive technical knowledge became scarce.[77]

Sources for black youths to learn about automobile technology prior to passage of the Smith-Hughes Act were already scarce. The spread of auto shop programs improved the situation but did not level the playing field. As early as 1910, Armstrong Manual Training School in the Washington, D.C., public school system offered a course in machine shop and minor repair work for young men aiming to become chauffeurs. By 1916, the year before the Smith-Hughes Act, black public four-year secondary schools that offered industrial training tended toward woodworking, tailoring, shoe repairing, and other traditionally black occupations. Some began offering other areas such as mechanical drawing, pipefitting, and metalworking, but automobile curricula such as Armstrong's remained rare.[78]

As World War I drew to a close, an aspiring black mechanic who wanted a structured automotive education would have had to look to the black colleges. Those colleges that had participated in the University of Uncle Sam were particularly quick to turn their experience toward training black mechanics and, eventually, black auto shop teachers. The Agricultural, Normal, and Industrial school at Greensboro, North Carolina, where the army had conducted a driving school for black troops, placed a long-running series of ads in NAACP cofounder W.E.B. DuBois's magazine of black culture, the *Crisis*. The first appeared in August 1919, a month before the signing of the Armistice, "offer[ing] to the Negro youth of the state opportunities that none can afford to neglect." Along with blacksmithing, numerous building trades, and "broom-making," Greensboro students could now enroll in auto mechanics courses. Prairie View State Normal and Industrial School in Texas followed suit. Having trained black mechanics in the University of Uncle Sam, the school turned its attention and equipment to training black mechanics along with numerous other trades. The work of providing blacks access to automotive technological knowledge, begun by entrepreneurs such as Lee Pollard and Ben Thomas a decade earlier, began finding support at black colleges even before most states' school systems began offering automobile courses in their high schools.

As federal Smith-Hughes money began flowing into public school systems across the United States in the 1920s, African Americans had, out of necessity, already begun building a separate system of automobile service and repair. Many whites did not easily accept black automobile ownership, neither north or south of the Mason-Dixon Line. Thus, black motorists throughout the first half of the twentieth century did not experience the "open road" in the way white Americans did. At every turn they faced segregation, discrimination, and potential violence. Many white-owned restaurants, hotels, filling stations, and garages would not serve black motorists. Even the National Park Service conceded that black motorists should be discouraged from entering national parks given that the hotels therein would not serve them. A middle-class black professional who wanted to use an automobile to take in the natural beauty of the West in the 1920s and 1930s would have found it difficult to obtain supplies and accommodations along the way.[79] The particular freedoms that automobility offered to African Americans, combined with white limits imposed on those freedoms, led African Americans to create their own network of businesses to support black motoring.[80]

By the late 1920s the black automobile economy accounted for a substantial portion of black-owned businesses and black employment. J. H. Harmon, studying black businessmen in the late 1920s, found that garage and service station owners ranked third in number behind grocers and "cleaners and shoe repairers." "Many Negroes," he observed, "while working for white stations have conceived the idea that such promising enterprises among their own people will pay enormous sums, and for the most part they have succeeded."[81] When the U.S. Census conducted the first nationwide survey of black-owned retail businesses in 1929, automobile-related sales and service represented the fourth largest generator of sales, behind food, restaurants, and "all other retail stores." This despite the fact that no African-American entrepreneurs could secure new car dealerships from major manufacturers, so revenues from black purchases of new cars went almost entirely to white-owned dealerships.[82] Filling stations, garages, and repair shops accounted for 91 percent of the 1,679 black-owned automotive businesses counted by the census in 1929. Proprietors and partners running those businesses numbered 1,707, and they employed another 827 workers.[83]

A year earlier a survey conducted by the National Negro Business League provided some insight into the rootedness of the black automobile economy in local black communities. Of the 169 automobile businesses included in the survey, all but one were black-owned, and none employed any whites.[84] Blacks could and did cross the problematic "color line" to work in white-owned shops, though usually as mechanics' helpers, greasers, or "wash boys." The reverse did not hold true in any of the surveyed black-owned shops: whites did not work in black shops. Located overwhelmingly in "colored neighborhoods," black-owned shops drew most of their customers from the immediate vicinity. Indeed, 15 percent of the shops reported no white customers, and 73 percent reported that 70 percent or more of their customers were black. Yet 42 percent of the black-owned automobile businesses reported that 30 percent or more of their customers were white.[85]

This white patronage of black automobile businesses in black neighborhoods exemplifies the permeable color line, or the "muddled middle," of early-twentieth-century segregation in the American South, where blacks and whites mixed more easily than sometimes supposed.[86] The ambiguous status of auto mechanics generally—at the same time servants and technological experts—no doubt contributed to this muddling of the color line. White patrons comfortable with their own status could patronize black mechanics as servants without threat to their own sense of white racial identity. Crossing the line in the other direction could be more problematic. Sociologist John Dollard noted in the mid-1930s that many whites grew agitated when blacks purchased better automobiles.[87] Consequently, blacks could be less certain of getting fair service in white shops. When black Alabama sharecropper Nate Shaw purchased a new Ford in the 1920s, he discovered on the drive home that the brakes had been assembled incorrectly by the white mechanics at the dealership.[88] Given the very common and routinized assembly procedures developed by Ford for assembling cars at the dealerships, it seems unlikely that the mechanics would have made such a mistake unintentionally. The promises of automobility in the face of segregation, the increasing purchasing power of blacks, and the antipathy of some whites toward black motorists fed the creation of a black automobile economy. Black business directories, newspapers, and other publications regularly carried ads for these businesses, and they filled a need in the black community which white firms could not or would not meet.[89]

The introduction of auto shop into public high schools improved young African Americans' access to the automotive technological knowledge needed to support this black auto economy, but unevenly (see fig. 17). In part this was due to geography. For both blacks and whites living in southern states, as well as for those living in northern and western rural states, the significantly lower spending on public vocational trade and industrial education limited their access to auto shop courses. To illustrate, in 1937 Mississippi sought a total of $4,608.45 in federal matching funds for all of its vocational education programs (which did not include any auto shop courses, black or white). By comparison, Minnesota in that same year accounted for $46,830.26 in federal matching funds for its vocational education programs—$3,015.72 of which supported 179 students in the all-day auto mechanics course at the Minneapolis Boys' Vocational School.[90]

In states with segregated public schools, high school auto shop grew predictably separate and unequal. The Smith-Hughes Act imposed no requirements on states to distribute funds equally between black and white schools, which led to great disparities in spending, particularly in the area of trade and industrial education.[91] One historical study found that "more than half of the federal funds which the southern states received for blacks on the basis of their population were in most cases consistently diverted to white schools."[92]

When states with segregated schools did offer auto courses, black students often faced limited curriculum and funding. During the 1929–1930 school year thirty white male students attended a three-year, all-day auto mechanics course at Little Rock High School in Arkansas—the same school system that would make national headlines in 1957 when President Dwight D. Eisenhower sent in the 101st Airborne to maintain order in the newly integrated high school. In 1930 Little Rock's auto shop program included shop work, three years of "related science," and one year of "garage management." That same year nine black male students enrolled in a two-year, all-day program at Pine Bluff, Arkansas, which taught only auto mechanics and "related work"—no garage management and not even the scant "related science" typical of most auto shop programs. Little Rock's white auto shop received $1,270.79 in federal matching funds, while Pine Bluff's program received $78.86. Little Rock's prospective white mechanics benefited from a social and political system that was willing to fund their training at nearly five times the per-student rate of prospective black mechanics.[93]

Nonetheless, African-American boys did enroll in high school auto shop when it was available. The actual numbers of these enrollments is unknown because only those states maintaining legally segregated schools were required to report separate enrollment numbers for black students in federally funded programs.

Thus, the number of black auto shop students in the large Chicago, New York, and Los Angeles school systems is unknown. Available information, however, does seem to indicate large black enrollments in auto shop where the opportunity was available, especially when compared to other trades such as shoe repair or janitorial work. Black enrollments also seem to have shifted between 1920 and 1937 from evening classes to all-day programs, indicating increased expectations by students and educators that graduates could anticipate moving into automobile-related jobs upon graduation. Still, in many cases the funding and curriculum of segregated auto shop programs prepared young black men to enter a world of work in which whites viewed them as porters, chauffeurs, greasers, and only rarely as mechanics.[94] These students' own expectations upon graduation may have been to work as mechanics or entrepreneurs of various auto-related businesses in the separate black economy. In either case ostensibly democratic public education in America supported and helped perpetuate a racially segregated automobile culture while at the same time increasing the gender segregation of general automotive knowledge and making auto shop and auto repair the domain of the white working-class male.

Auto shop rapidly became an institution within American public education,[95] serving the needs of motorists and manufacturers for a permanent class of technical service workers to support mass automobility. It also met the cravings of some adolescents to learn about and master a technology that was quickly becoming associated with masculinity, independence, freedom, and adulthood. Auto shop programs, however, did not flood the labor market with waves of experienced mechanics following each June graduation. Auto shop programs introduced students to the principles of automotive technology and basic repair techniques while socializing them to the expectations of employers. None could truly replicate the knowledge gained by years of experience with employers, fellow mechanics, customers, and their cars in a commercial shop setting. Such was not the goal of educators. Rather, auto shop sorted and fed adolescent boys to the automotive service industry, where they could be further trained in particular methods and procedures by their employers.[96] By the mid-1930s a relatively stable system emerged for creating generations of automobile mechanics through a combination of high school auto shop, on-the-job experience, and continued training. This system, and the class, gender, and race associations that it embodied and perpetuated, remained essentially unchanged for most of the remainder of the twentieth century.

Fig. 1. Liveried coachmen for New York's elite families awaiting orders from their masters. The Vanderbilt-Marlborough Wedding, St. Thomas Church, Fifth Avenue and 53rd Street, 1895. Courtesy of the Museum of the City of New York, Byron Collection.

STEVENS-DURYEA TOURING CAR WITH STANDARD EQUIPMENT
Stevens-Duryea Co., Chicopee Falls, Mass.

Fig. 2. Stevens-Duryea Touring Car, 1910. The Stevens-Duryea was typical of the chauffeur-driven cars of the time. Note the toolbox on the running board for easy access. The cylinder tank at the front of the running board contained acetylene gas for the lamps. Reprinted from Charles B. Hayward et al., *Cyclopedia of Automobile Engineering,* vol. 1 (Chicago: American Technical Society, 1910), frontispiece.

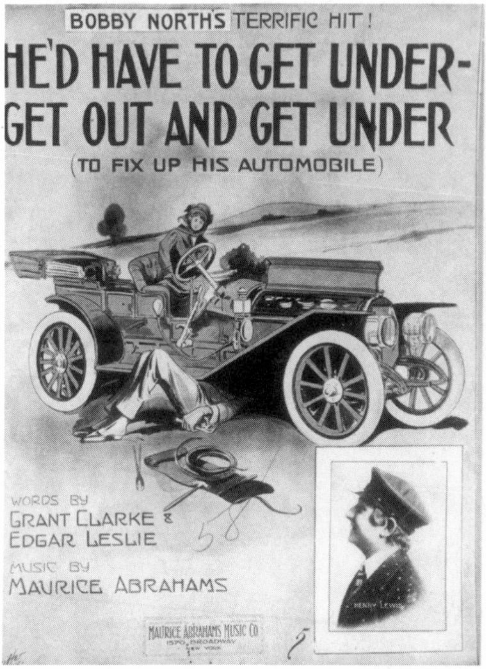

Fig. 3. Sheet music for a popular song in 1913 reflecting the temperamental nature of early touring cars. Note the tool pouch and coil of wire in the foreground. Courtesy of the University of Delaware Library, Newark.

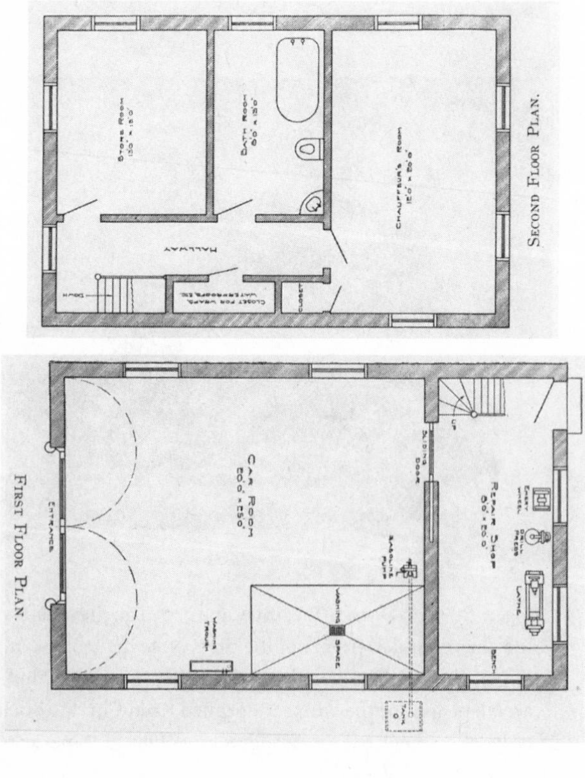

Fig. 4 Floor plan of an ideal private garage, with the chauffeur's living quarters on the second floor. The equipment in the repair shop includes an "emery wheel," or grinder, a drill press, and a metal lathe. Because replacement parts were not readily available, chauffeur-mechanics had to fabricate or repair many parts in order to keep the automobile in operation. Reprinted from "A Model Garage," *Horseless Age* 17, 30 May 1906, 819.

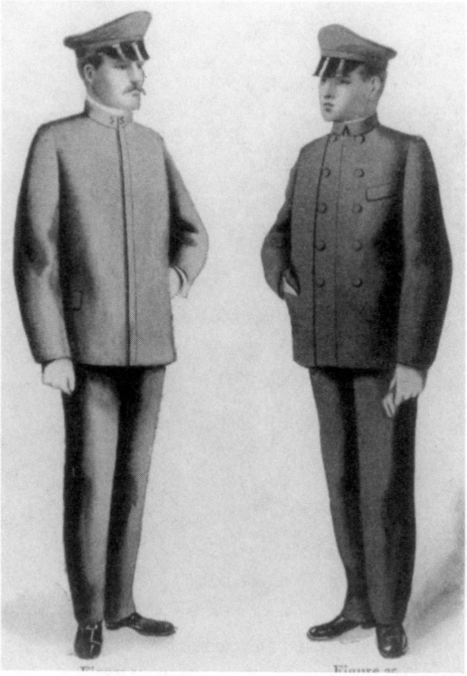

Fig. 5. Chauffeurs' livery offered by Saks and Company of New York, 1903. Saks and Company, *"Automobile Garments and Requisites, for Men and Women,"* New York, 1903, 40–41. Courtesy of the Hagley Museum and Library, Trade Catalog Collection.

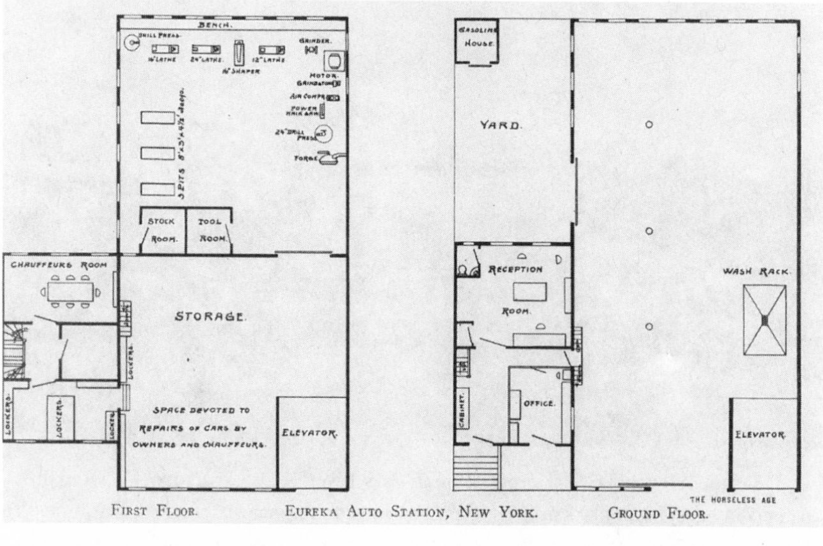

Fig. 6. The Eureka Auto Station was on West 124th Street in New York City. Note the "Chauffeurs' Room," the "Space Devoted to Repairs of Cars by Owners and Chauffeurs," and the many machine tools, including three lathes, a shaper, a press, and a forge. Reprinted from *Horseless Age* 23, 21 April 1909, 512–22.

Fig. 7. Eager young entrepreneurs offering sporting goods, bicycle repairs, and auto repairs in Burbank, Calif., 1908. Courtesy of the Security Pacific Collection, Los Angeles Public Library.

Fig. 8. Samuel Holland's Blacksmith Shop in Park River, N.Dak., ca. 1910, showing the diversity of equipment blacksmiths might be called on to repair in rural communities. To the far left and right sit steam tractors. In the foreground three men hold wire bicycle-style wheels with pneumatic tires. In the center middle ground sits a three-wheeled motorcycle, and behind it are three women sitting atop a horseless carriage. Courtesy of Fred Hulstrand History in Pictures Collection, North Dakota Institute for Regional Studies, North Dakota State University, Fargo.

The Modern Village Blacksmith. —From The New York World.

Fig. 9. A 1915 cartoon depicting the frustration some blacksmiths felt when working on tourists' automobiles. The fashionably attired motorist and his wife sit idly by while the blacksmith drips sweat from his brow and his fire burns unattended in the shop. The road sign in the distance indicates the way to New York. Reprinted from *Blacksmith and Wheelwright,* November 1915, 865.

Fig. 10. Unidentified early auto mechanics, ca. 1910. Note the majority of eager young faces and the confident posturing about the partly disassembled automobile. Anonymous, undated photograph from author's collection.

Fig. 11. Advertising brochure for New York's West Side YMCA Automobile School, 1909–10 season. Courtesy of the Warshaw Collection of Business Americana—Automobile Industry, Archives Center, National Museum of American History, Behring Center, Smithsonian Institution.

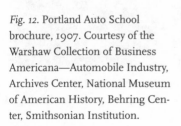

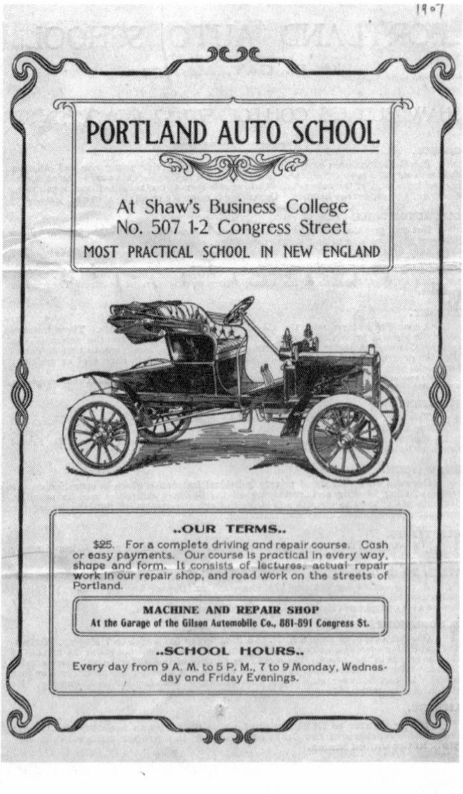

Fig. 12. Portland Auto School brochure, 1907. Courtesy of the Warshaw Collection of Business Americana—Automobile Industry, Archives Center, National Museum of American History, Behring Center, Smithsonian Institution.

Fig. 13. Motor Transportation Corps full-page recruiting advertisement seeking "Gas Hounds," 1918. Reprinted from *New York Times,* 3 November 1918.

Fig. 14. Auto shop class at Boys' Technical High School of Milwaukee, 1928. Graduates received a "special Trade Diploma" indicating they had "completed a course of training in his chosen field" but which did not qualify them for college admission. Reprinted from Harry W. Paine, "A Survey of the Boys' Technical High School of Milwaukee, Wisconsin, and the Organization of Its Automotive Department" (master's thesis, Iowa State College, 1928), 14, 26.

Fig. 15. U.S. Office of Education photo of auto shop students in Battle Creek, Mich., public school, 1940. Reprinted from Negative no. 12-E-39-884, Record Group 12-E, Records of the Office of Education, Prints: Education and Training, 1936–48, National Archives, College Park, Md.

Fig. 16. U.S. Office of Education photo of electrical shop students in Battle Creek, Mich., public school, 1940. Reprinted from Negative no. 12-E-39-886, Record Group 12-E, Records of the Office of Education, Prints: Education and Training, 1936–48, National Archives, College Park, Md.

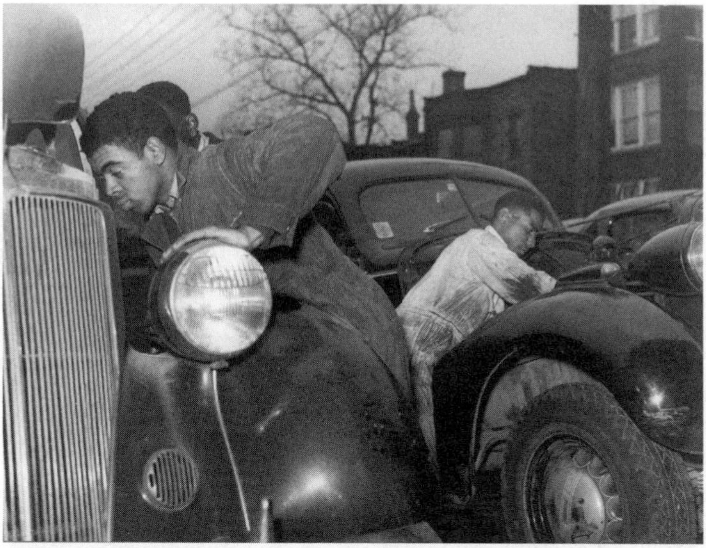

Fig. 17. African-American auto repair students at Drexel Institute of Technology, Philadelphia, 1942. Reprinted from Negative no. 12-E-39-3348, Record Group 12-E, Records of the Office of Education, Prints: Education and Training, 1936–38, National Archives, College Park, Md.

Fig. 18. Auto mechanic Lynn Reynolds and partner in front of their small shop in Mason City, Iowa, in the 1930s. Shops such as this were numerous and would not likely have imposed strict flat rate and piece rate pay systems on themselves or their few employees. Photo courtesy of Scott Sandage.

Fig. 19. Promotional image of the Ford Laboratory Test Set being used by a mechanic to convince the customer of needed service, 1930s. From the Collections of The Henry Ford Museum.

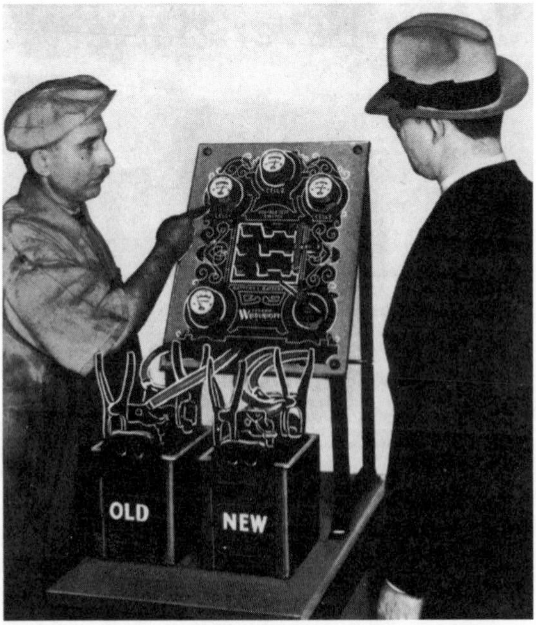

Fig. 20. Weidenhoff Star Battery Seller, 1936. Reprinted from Joseph Weidenhoff, Inc., *Catalog No. 45*, 1 January 1936. Courtesy of Smithsonian Institution Libraries, Washington, D.C.

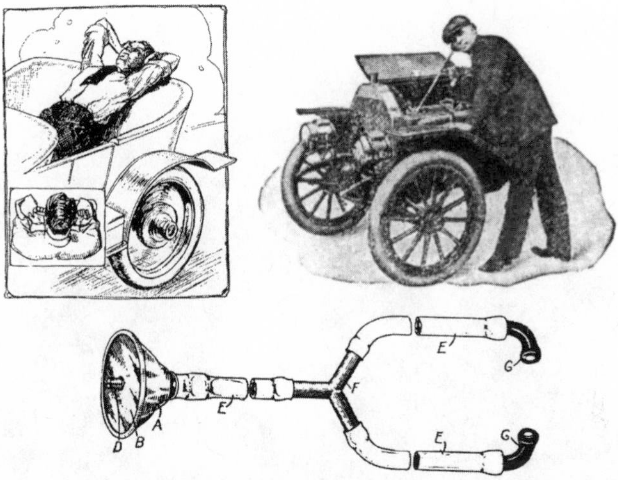

Fig. 21. Various methods of listening for trouble. The top left image instructed mechanics how to listen for rear axle noises. The top right image, demonstrating the use of a "sounding rod" to diagnose engine knocks, persisted through more than twenty editions of the Dyke's widely consulted reference book—despite the poor quality of the image and the increasingly out-of-date automobile shown. The bottom image provided a plan for a "cheap but practical sound locator or stethoscope." Reprinted from *Dyke's Automobile and Gasoline Engine Encyclopedia,* 6th ed. (St. Louis: A. L. Dyke, 1918), 638, 739; and "Locating Automobile Sounds," *Automobile Dealer and Repairer* 15, July 1913, 49.

Fig. 22. Mother and daughter mechanics, created by acclaimed illustrator Al Parker. During World War II, even women who did not don overalls professionally were encouraged to do so at home as part of their domestic contribution to the war effort. Cover art reprinted from *Ladies' Home Journal,* March 1943. Image © Kit Parker.

Fig. 23. A group of boys under the hood of a car, 1951. Courtesy of A. Y. Owen, Time and Life Pictures, Getty Images.

Fig. 24. Smokey Yunick using a screwdriver as a sounding rod to listen to a race engine. Reprinted from *Popular Science,* January 1964, 94.

Fig. 25. Cover of *Ford Service Life* magazine for dealership mechanics, May–June 1971, depicting the masculine and heroic image of the race car mechanic, which starkly contrasted with many mechanics' day-to-day experience. Courtesy of Ford Motor Company Archives.

Fig. 26. Unsavory image of auto mechanic. Reprinted from Roger Riis and John Patric, *Repair Men May Gyp You* (1949), frontispiece.

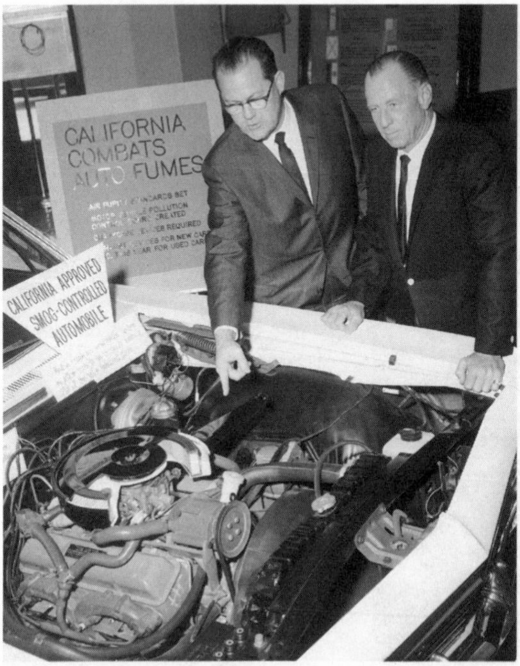

Fig. 27. Automobile Club of Southern California executive vice president, Joe Havenner (left), points to belt-driven compressor of emission control system on a 1966 Chevrolet. Hoses leading over valve covers, down toward exhaust manifolds, carry additional oxygen to exhaust stream. Photo courtesy of Automobile Club of Southern California Archives.

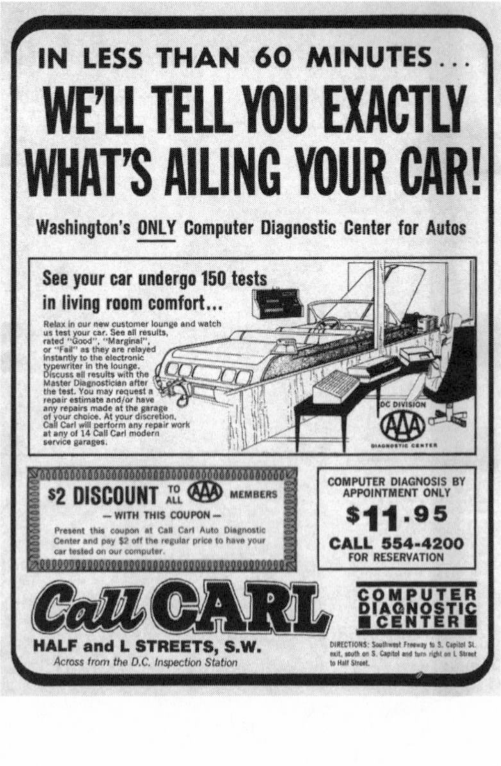

Fig. 28. "Call Carl's Computer Diagnostic Center" ad, *American Motorist,* December 1967. Courtesy of Call Carl Files, Transportation Collection, National Museum of American History, Behring Center, Smithsonian Institution.

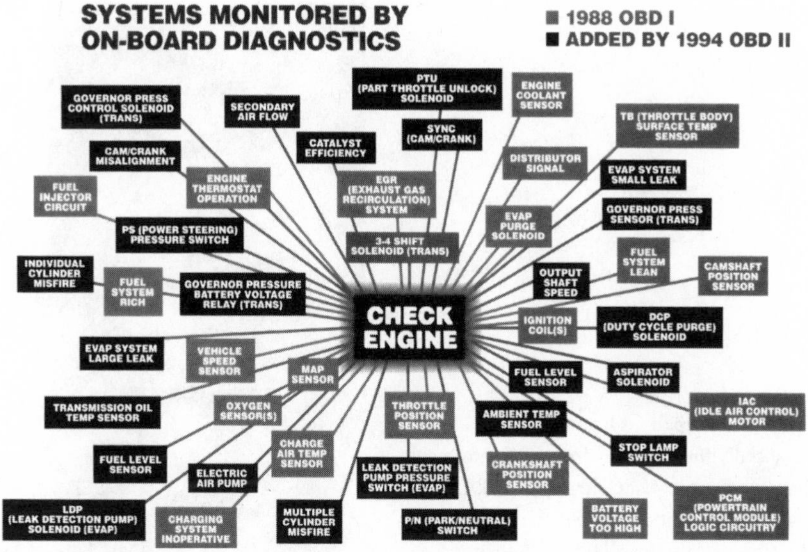

Fig. 29. Diagram of system sensors integrated into OBD and OBD-II systems and the "Check Engine" light. Reprinted from *Car and Driver,* December 1995, 88.

Fig. 30. Engine compartment of 1999 Volkswagen Passat showing plastic demi-hood obscuring the mechanical details of the engine. Photo by author.

Fig. 31. Auto repair ad from 1930s depicting the socially incongruous image of auto mechanics as surgeons. Courtesy of Call Carl Files, Transportation Collection, National Museum of American History, Behring Center, Smithsonian Institution.

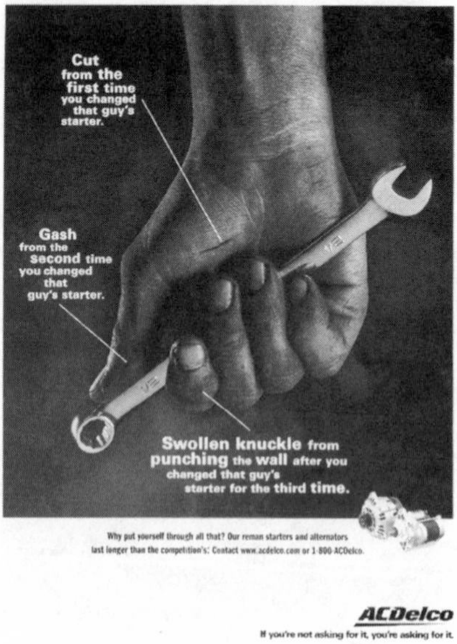

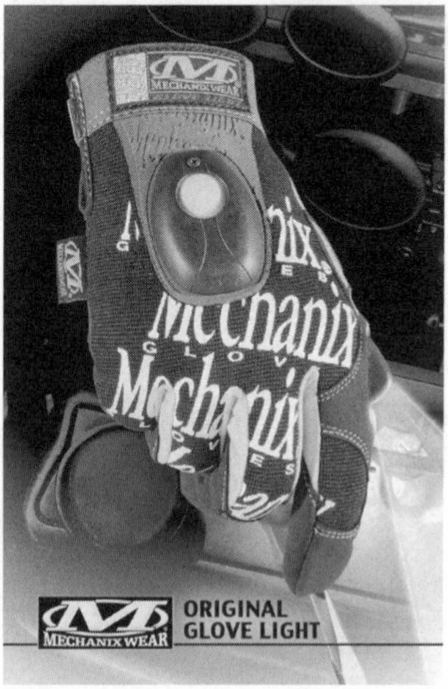

Fig. 32. Advertisers still use the blue-collar masculine image of the strong, dirty, calloused hand to sell their product, but increasing numbers of mechanics now prefer to protect their hands. (Mechanix Wear's Glove Light has a built-in 10,000 lumen LED light.) Image on left courtesy of AC Delco. Image on right courtesy of Mechanix Wear.

CHAPTER 5

Tinkering with Sociotechnical Hierarchies

Creating new mechanics through auto shop helped resolve the problem of determining who would be responsible for maintaining the millions of new automobiles being sold every year. Motorists no longer had to employ chauffeur-mechanics, wait at blacksmiths' shops, or seek out machinists or plumbers. They could now turn to automobile mechanics trained to make full-time careers out of maintaining and repairing their cars. Yet creating and training auto mechanics did not resolve "the service problem." Education did not noticeably improve or ease the anxiety that motorists felt when they entered the repair shop.

Rather, the system for creating mechanics which evolved out of the 1920s reinforced the class and gender disparity between motorists and mechanics by locating crucial technological knowledge within a narrow demographic band. In this context gender, class, and sometimes race became almost universal shorthand among mechanics for technological ignorance. Mechanics almost never saw a girl in their auto shop classes or a female mechanic in their workshops, in their factory seminars, or in their industry trade literature. Women had been socially and institutionally barred from automotive technical knowledge, thus many mechanics assumed they were dealing with ignorant motorists when they dealt with female customers. Similarly, in the eyes of many mechanics wealthy male motorists as well as African-American motorists were no more likely to understand their automobiles than women. In a shop culture that valued masculine working-class traits such as strength, physicality, and mechanical skill, such technical ignorance did not curry respect from mechanics.[1] In response, mechanics could, and did, use their technological knowledge as leverage in service relationships, working faster or slower, being polite or impolite, pleasant or unpleasant, based on their own perceptions of opportunity and justice in each exchange relationship. Conversely, because of the way the mechanic's knowledge and occupation had been socially defined and devalued, customers and employers often did not treat mechanics with the respect these tradesmen felt they deserved.

Together mechanics and motorists perpetuated the cycle of distrust and anxiety at the root of the service problem, leading manufacturers, dealers, and others in the automotive service industry to attempt other reforms beyond creating more mechanics. They sought to ease tensions either by physically altering automotive technology or by changing the public perception of automobile mechanics and the repair process. In the words of one industry reformer, their intention was "to change the mental picture carried by the average car-owner."[2]

Re-creating Mechanics as Licensed Professionals

Predating the public school auto shop movement, some participants in the early automotive service industry sought licensing or certification as means to reform the auto repair trade. Given the diverse backgrounds and abilities of ad hoc mechanics and the multiplicity of private automobile schools and training programs for aspirants, determining who could be trusted to do reliable work was probably more difficult in the years around World War I than at any time before or since. Responding to consumer complaints about the incompetence of many garages and mechanics, legislators and industry observers reasoned that automobile mechanics or garages should be licensed or certified in some fashion.

Writers in the automotive press had toyed with the idea before the war and generally favored some means of either certifying or licensing mechanics. The Wisconsin Automobile Business Association of Milwaukee in 1916 "found considerable sentiment in favor of the establishment of a system of licensing repairmen or journeymen employed in garages."[3] A writer for *Motor World* believed, naively in retrospect, that such a system would enjoy the support of a broad coalition of motorists, mechanics, and employers. The war emergency diverted attention from the problem for a few years, then in the summer of 1919 the state of Oregon announced a plan for statewide exams and licenses for mechanics. Later the same year Frederick G. Bagley, commissioner of public affairs in Buffalo, New York, outlined his plan to license all mechanics in that city, including "an examination to determine their fitness." Bagley believed it quite reasonable to license mechanics because, he argued, "even barbers must prove their fitness to practice their profession in most cities."[4] Some employers, such as larger automobile dealers, initially agreed, seeing examination-based certification as a way to help find qualified employees. Thus, by the late 1910s various state and local governments began formulating legislation to establish reliable indicators of mechanical competence through automobile mechanic licensing bureaus.[5]

In practice licensing laws surprised some in the industry by providing a rallying point around which to organize a dispersed and previously unorganized labor force. Implementation of Oregon's law required mechanics in that state to cough up a five-dollar fee to cover program expenses or to pay a one hundred–dollar fine for practicing without a license. In response, at least some Oregon mechanics became unified and vocal in their opposition to the plan. Mechanics in Portland threatened to walk out over the issue, and service managers and automobile dealers nationwide took notice.[6]

With similar legislation pending in various states and localities across the United States, service managers, automobile dealers, and their trade press allies reconsidered their support of certification. Mechanics in nearby Seattle and Tacoma, Washington, had previously established a self-certification process through their strong union organizations in those cities and commanded higher wages.[7] The growing association between licensing/certification and unionization troubled some important opinion shapers in the strike-plagued postwar years. Employers and editors began to question whether the costs of such plans—either in terms of sparking the unionization of mechanics or in terms of increased wage expectations of licensed mechanics—outweighed the benefits. The editor of *Motor Age* cautioned that the Oregon law had made "organization of the mechanics into a labor union an exceedingly simple matter . . . motor car dealers and repair shop owners must decide whether they wish to support or combat such legislation."[8] *Motor World,* commenting on similar bills pending in nine states, wrote, "Legislation such as is proposed in these states contains a menace to the industry and should be thoroughly studied by dealers' and garagemens' associations whenever it comes up."[9] Predictably, employers bolted from their tenuous coalition with consumers and quickly doused the nascent licensing movement.[10]

Recreating the public image of mechanics through licensing or otherwise granting them quasi-professional status became a dead issue by the mid-1920s. Licensing resurfaced briefly during the Great Depression as a strategy to prevent unemployed, itinerant, and "gyp" mechanics from siphoning business away from garages. Fear of unionization and increased wages remained a strong deterrent to employers, however, and no legislation materialized. Consequently, automobile mechanics lacked one of the basic hallmarks of professional status for most of the remainder of the twentieth century.[11]

Reconfiguring Automobiles

The service problem had a greater effect on the shape of automotive technology during the 1920s. Automakers had not ignored the service problem earlier, but their engineering concerns were oriented toward production and toward designing various parts so that they could be assembled at lowest cost. To the degree that they considered service, they focused more on improving reliability, which meant designing parts that lasted longer. It meant experimenting with various designs for cylinder castings in an attempt to achieve maximum cooling and strength without cracking during long-term use. It meant experimenting with metals and designs for ball bearings, thrust bearings, and bushings. At Ford it meant using vanadium steel to produce tougher front axles and gears for the Model T.[12]

These concerns with production and reliability continued, but during the early 1920s the accessibility of parts requiring regular adjustment or repair received increased attention from design engineers. By that time the overall form of automotive technology had essentially stabilized—with a gasoline engine in front, a transmission in the middle, and drive wheels in the rear—so that more attention could be devoted to the specific arrangement of parts within that basic form. Furthermore, the rising importance of service to new car sales during the 1920s gave such concerns higher priority in the design process. An increasing number of customers were not first-time car buyers; during that decade many were buying new cars as replacements for their older models. Previous experiences had taught many drivers to be more cautious in their next purchase: they wanted neither to perform nor pay for service operations that required extensive disassembly of their vehicles to gain access to minor parts. Thus, in the 1920s automakers and their design engineers turned their attention to the topic of "accessibility."

T. F. Cullen, managing editor of *Automobile Trade Topics,* criticized automakers' failure to address service concerns in their design. Speaking to a meeting of the Society of Automotive Engineers (SAE) in early 1921, Cullen summarized his survey of forty makes and models of cars and rated them "according to their comparative accessibility for 15 typical service operations." He concluded that there was "considerable room for improvement in service accessibility" and illustrated examples of good practice that he hoped other automakers would follow. B. M. Ikert, editor of *Motor Age,* addressed the SAE the next year and chastised his audience by citing more than two dozen specific examples of poor design from the service man's point of view. On one model car the mechanic had to pry the body up to remove the battery. On another, one had to remove the radiator to replace a fan belt. Another was built so that one had to remove the running board in order to remove the bolt holding the commonly cracked or broken spring. In his own journal Ikert balanced his criticism by praising designs that improved service accessibility.[13]

A chorus of voices joined Cullen and Ikert in raising concerns about accessibility. Improving accessibility, according to some, would improve the service relationship by reducing the need for mechanics to explain large labor bills when making otherwise simple repairs. As P. J. Durham of the Automotive Electric Service Association put it: "It is ludicrous to see a husky mechanic, lying under a cowl-board, trying with his No. 9 hands to attach a wire under the head of a 6-32 screw. But, it is not funny when an owner objects to a large labor-charge for electrical service-work that is inescapably long because overgrown watch-screws must be fiddled with. It is very bad."[14]

Voices representing service men suggested that the best way for design engineers to understand service needs would be for them to get out of the office more often and visit the service shops of their employer's dealers and branches. As A. J. Cook put it, "If some of the engineers would put on overalls and get underneath the car as the 'grease balls' do, they would be in closer touch with the details that give trouble to the service-station." Indeed, some automakers used roadmen who visited dealer shops to provide factory service information while carrying back concerns from the field. Others provided forms for service managers to report continuing service problems related to design. Still others established their own service-testing facilities to examine cars in service under controlled conditions.[15] Service concerns clearly affected at least some of the internal configurations and shape of automobiles in the 1920s.

For the most part, however, engineers seemed unwilling to be tutored on design issues by lowly shop mechanics. The growing social chasm between auto mechanics and automotive engineers adversely affected the manner in which automakers gathered information regarding service needs and thus how they designed their cars. The responses of some engineers to suggestions from service men indicate that an uneasy relationship may have hindered the flow of information. As one commentary put it, "Certainly nobody with an appreciation of a chief engineer's responsibility and the value of his time would expect him to don overalls daily and assume the role of a grimy mechanic." Society of Automotive Engineers president B. B. Bachman answered Ikert's critical assessment of design engineers by claiming "It is not necessary for the man doing service work to tell the engineer the trouble and how to fix it. . . . The man doing service work has difficulty enough to fix troubles, and I believe that he is not usually possessed of the necessary qualifications to enable him to analyze, and suggest remedies." During the same SAE meeting J. C. Gorey remarked: "If the man who is in the field to repair and remodel different units would stop and consider the wonderful growth of this industry, he would try to remedy some of the troubles and lessen the number of complaints with which he is annoying the engineering department. . . . Let the engineer fathom the greatest trouble and let everybody else do his part."[16] Instead of listening to mechanics in the shop, engineers sought to teach them a thing or two about repairing cars with system and dispatch.

Re-creating Mechanics as Rationalized Workers

Factory engineers and garage service managers wanted to reform the motorist-mechanic relationship as well as the mechanic-employer relationship by establishing standard procedures, times, and charges for specific repairs while gaining tighter managerial control over mechanics. Improvements in service accessibility resulted as much from automakers' drive to impose flat rates as they did from any compassionate concern for the troubles of the "grimy mechanic."[17] Through the establishment of flat rates, automakers, dealers, and some independent garages attempted to recast the public image of automobile service as something reliable, predictable, rational, and safe, rather than anxiety inducing. Just as a customer could go into a store and buy a suit or shirt in the size needed at a set price and leave, flat rate advocates believed motorists should be able to go into a garage, get the repair or service they needed for a set price, and leave without feeling "gypped."[18]

Closely related to the scientific management movement that had profoundly affected production industries, the flat rate movement attempted to determine the time it "should" take to perform specific repair operations, such as replacing the brake linings on a 1920 Hudson. Some large shops established their own flat rates by simply tracking and averaging the time mechanics spent on specific types of jobs. The Ford Motor Company established flat rates for its dealers through elaborate time-motion studies in controlled settings. These studies in turn highlighted the need to improve service accessibility in the design stage. By referring to a compilation of such flat rates, a service manager could quote the customer a "fair" and "accurate" price for a job up front, before beginning work on the car. Employers then expected mechanics to perform the repair in flat rate time or less. They hoped to speed up mechanics' work by instituting piecework wages along with flat rate pricing. In such a system a mechanic would receive his regular rate of pay for the time allowed a given job in the flat rate charts, whether it took him less time or more. If he worked faster, he would effectively increase his hourly rate of pay. If he worked more slowly, his paycheck would suffer.[19]

The flat rate / piecework movement of the 1920s meant more, however, than controlling labor costs. It was also an attempt by some employers to re-create mechanics as passive, interchangeable workers who performed discrete repair operations according to prescribed methods. In this sense the flat rate movement represented the antithesis of the short-lived licensing movement and its implied professionalization of the trade. The president of the Automotive Service Association of New York declared, "System and short cuts in repairings [*sic*] must be developed, outlining the most efficient plan possible for our men to follow instead of allowing them to outline their own."[20] Various coalitions of employers, service managers, and factory engineers collaborated to determine the "proper" methods for specific repairs and established flat rate times based on those methods. In some instances each step of a repair operation was written out in detail to instruct the mechanic in the approved method. Williams and Hastings, a large Detroit dealership, developed a series of "operation sheets" providing detailed instructions on "the best methods and the best routine or sequence of operations to do the work properly and with the least effort."[21] The Franklin Motor Company's flat rate data filled seven volumes in 1925, four of which provided detailed "layouts" of the sequences of labor operations, time allowances, and special tools and supplies needed for each flat rate labor operation.[22]

Not all flat rate systems provided such detailed instructions to the mechanic, but by rationalizing and codifying automotive repair knowledge, each aimed at removing control of that knowledge from mechanics and placing it in the hands of employers and service managers. Ideally, many of these simplified and rationalized repair procedures could be carried out by less experienced (i.e., lower-wage) mechanics coming out of the public school auto shops, who would be more tractable than experienced mechanics. According to Don Hastings, vice president of Williams and Hastings, if a mechanic bucked the flat rate system, "we simply get rid of him."[23]

Yet Hastings likely did not have such absolute power to impose reforms in his shop as he wanted his audience to believe. Even the Ford Motor Company backed off its advocacy of flat rate / piecework reforms by the early 1930s, due in large part to the opposition of mechanics. Mechanics resisted the imposition of flat rate / piecework reforms most directly by quitting and moving to other shops or leaving the trade altogether. One-third of Williams and Hastings's mechanics quit within a week of adopting the system.[24] Mechanics elsewhere restricted output, haggled over time allotments, or performed fast but shoddy work. In some instances putting mechanics on piecework pay exacerbated tensions with customers. Many mechanics refused to do any work beyond what was on the repair order and included in the flat rate operations. If a carburetor needed adjusting to make a fresh valve job run sweetly, it would be left undone unless specified and included in the mechanic's pay. Customers who wanted to chat with or watch the mechanic while he worked often got brusque or surly treatment, and if a service manager was not able to greet a customer quickly, piece rate mechanics did not see it as their job to greet the customer at all.[25]

Flat rate / piecework reforms also failed to resolve the service problem because those in charge of implementing the system saw mechanics as a large part of the problem. Meanwhile, smaller garages and independent mechanics did not see themselves as part of a service problem and ignored the system. Owners of these small shops were usually experienced mechanics who worked alongside their employees. They knew what each mechanic could and could not do. They did not need to rationalize or codify their mechanics' technological knowledge. These smaller shops could freely use published flat rate charts to aid them in giving customers more accurate estimates, but they did not generally adopt the contentious piecework plans.[26] In this way they could improve their relationships with customers at a time when large shops and dealers were discovering the drawbacks of strict flat rate / piecework plans. This may help explain why many customers preferred small shops to dealerships in the 1920s and 1930s (see fig. 18).

By the early 1930s it became apparent that the flat rate / piecework system would not dominate the auto repair industry, nor would it go completely away. In 1929 the Ford Motor Company, an early and strong advocate of flat rates and piecework pay, "reversed its position and [instructed] its dealers to abandon piecework."[27] A 1935 company publication advised that "the hourly rate basis of paying Dealers' mechanics has many advantages over the various commission [i.e., piecework] plans."[28] Over the course of a little more than a decade automakers, dealers, and service managers discovered that they could not easily impose production-derived management techniques on auto mechanics, yet the cost-controlling potential of the system remained attractive to some.[29] Mechanics of the period attempted to prevent the further degradation of their occupation by opposing piecework pay, yet the battle did little to raise their public esteem. Instead, the contemporary development of a stratified vocational education system in public schools had essentially locked in the public image of mechanics as underachieving working-class males.

Re-creating Mechanics as Laboratory Technicians

During the economic crises of the 1930s service gained new importance in the eyes of manufacturers and dealers. With the slump in new car sales brought on by the Great Depression, automakers and dealers looked anew to service as a vital source of revenue. They sought to increase their service business but recognized that customers struggling to pay for food and shelter were skeptical of mechanics who tried to convince them that their cars needed service. A gulf of distrust and anxiety, which had grown out of society's attitudes toward the work of auto repair, separated customers and mechanics.

In the wake of failed efforts to rationalize and speed up repair work, a new wave of reform sought to "scientize" auto repair. At a time when World's Fairs promoted a bountiful and harmonious future made possible through advances in science, automakers, dealers, and equipment manufacturers developed and introduced large, scientific-looking diagnostic equipment in an attempt to bring the mantle of science into the repair shop. Just as popular media portrayed science as objective and authoritative, equipment makers intended these units to communicate objective facts about a car's condition with unimpeachable authority, thereby mediating between the mechanic and the customer and facilitating increased sales (see fig. 19). In typically American fashion designers of these diagnostic units turned to technology to resolve a social problem rooted in differing relationships to technology.

Diagnostic instruments such as the Ford Laboratory Test Set, the Stromberg Motoscope, and the Weidenhoff Motor Analyzer brought an assortment of existing tools and instruments together into a single, oversized rolling cabinet. These units sported impressive arrays of dials, gauges, switches, and chrome plating for the express purpose of selling more service. They could accomplish this, according to their makers, because although a customer may distrust the mechanic's assessment of his car's condition, he would trust the ostensibly objective opinion displayed on the dials and gauges of the diagnostic unit. The Ford Motor Company assured its dealers: "He [the customer] may doubt what your service man tells him. But he will be impressed by the scientific, visual evidence of the Ford Laboratory Test Set. . . . Frequently he will heed its warning, when he would be inclined to 'laugh off' the advice of the service man or procrastinate in accepting it."[30] Instrument makers created these diagnostic tools to circumvent tensions between customers and mechanics by displacing authority from the mechanic to an impressive-looking "instrument," making it easier for customers to accept and submit to the mechanic's prescription. Such units appeared rather suddenly in the 1930s, yet they embodied ideas and technology that had been circulating in the automotive service industry since at least the early 1920s. Instruments, mobility, and display had all been used previously in automobile service, but they had not yet been so forcefully combined.

Prominent features on the diagnostic units of the 1930s included large voltmeters, ammeters, and ohmmeters. Since as early as 1912, when Cadillac introduced the first production car with an electric starter, mechanics encountered problems with increasingly sophisticated electrical systems. Finding problems in such systems sometimes required the use of instruments to establish the integrity of circuits, motors, generators, and batteries. With few exceptions before 1930, however, these instruments were designed solely for the mechanic's use, when mechanics used them at all. The Ambu Electric Trouble Shooter, the Testall Trouble Finder, the Weston Fault Finder, and the Niehoff Defectometer were all small, portable units with rather businesslike appearances. They were small enough for a mechanic to place on a cowl, fender, or seat while in use and to store in a toolbox or bench otherwise.[31] For larger shops, or those that specialized in auto electric work, the Joseph Weidenhoff Company and the Paul G. Niehoff Company produced large, stationary test stands that employed large instrument panels similar to the 1930s diagnostic units. Such test benches, however, would have been installed deep within the shop and were only used on parts already removed from a customer's car.[32]

Some rolling instrument carts were also available in the 1920s. The Niehoff Company offered a low-slung electrical testing and charging unit called the "TestKart," about the size of an ice cream vendor's pushcart. Niehoff touted it as "an electric service station in itself" which enabled a mechanic "to test the electrical equipment of a car without bringing it into the service station." Among other things it could quickly remagnetize Ford magnetos and get customers' back on their way. Niehoff designed its TestKart to provide quick curbside service and convenience rather than building service income by establishing authority.[33] Along the same lines, Chevrolet's 1924 repair manual provided instructions for building a simple rolling electrical test stand with a voltmeter, ammeter, six-volt lamp, six-volt battery, test leads, and battery leads. Its design, too, was purely functional and was not intended to impress customers.[34] Even when the Kent-Moore tool company offered the same unit for sale in 1931, its aesthetics were little improved, and the company made no mention of using it as a sales tool.

Mechanics of the 1920s certainly knew how to use the power of display to convince motorists of needed repairs. While critics decried the junkyard appearance of many shops, the ability to show a customer a badly scored crank journal or a burned valve and valve seat favored keeping some "junk" demonstration parts around the shop. Even the ubiquitous inverted-piston ashtray could be a selling tool in the right situation. Mr. Bohler, the service manager of the Black and Maffett Dodge dealership in Atlanta, fashioned a demonstration stand for a partially disassembled, worn-out engine block. He could turn the engine to any angle to show customers the various internal parts of the engine and point out how lack of adjustments and poor lubrication could lead to increased wear and damage. Bohler found his unit to be "an invaluable aid in convincing the customer," and using it, he sold more service.[35]

As early as 1920, at least one electrical instrument maker sensed the marketing value of scientific-looking instruments. The company produced a small meter unit to be worn on the mechanic's wrist. Using two test leads, mechanics could perform various tests in view of the motorist while explaining the significance of the meter readings. "An instrument like this in the mechanic's hands used in the owner's presence produces a feeling of assurance in the owner's mind that the mechanic knows his business."[36]

Selling this "feeling of assurance" in service customers led equipment makers to combine instrumentation, mobility, and display in the diagnostic units of the 1930s. These units did not introduce significant new diagnostic capabilities. Rather, their makers designed them primarily for the visual consumption of motorists. The Joseph Weidenhoff Company said of its Motor Analyzer in 1936: "[It] has a direct merchandising appeal . . . [it is] handsomely finished . . . very impressive to [the] motor vehicle operator . . . [and] helps the shopman talk with unquestionable authority because it deals only with facts intelligently presented."[37] The Sun Manufacturing Company claimed: "It has a psychological effect on the customer when you take the Sun Motor Tester to his car and start checking the electrical and mechanical sections systematically. . . . [It] convinces the customer that you know what you are doing."[38] The Stromberg Motoscope Corporation bragged that the Model C-3 "has a great merchandising value as it combines showmanship with the utmost accuracy and utility," while its Model B unit "produces unquestionable results." Of their Mercury Vacameter, a manometer available separately or as an accessory on the Motoscope C-3, Stromberg pointed out that the gauge was visible from both sides of the unit: "This means that both the operator and the customer can see the actions of the mercury column as it rises and falls . . . It is a salesman in itself."[39] J. B. Elliott, a Ford dealer in Chadbourn, North Carolina, reported in 1935: "The Ford Laboratory Test Set has enabled us to sell more parts—and more labor with them. . . . Customers are drawn from a greater radius to get this service which we have advertised. . . .

[The Ford Laboratory Test Set] acts as the best salesman we have."[40] Motorists still had to rely on the mechanic to interpret the displays of the diagnostic unit because they no more understood the dials and gauges than they understood the automotive troubles that first brought them into the shop. Using the diagnostic unit, however, gave the mechanic's opinion new authority.

During the 1930s equipment makers also incorporated oversized display gauges into wheel alignment units and battery testers for the purpose of convincing customers to have service work performed. The mediating role they intended these units to fill is captured especially well in a sales image for Weidenhoff's Star Battery Seller (see fig. 20). In this image a dirty mechanic with strong ethnic features points to the "highly polished copper plated" gauges of the Battery Seller, while a tall Anglo-Saxon customer in business attire looks on. It projects the clear message that the gauges on the Battery Seller would bridge the social gap between the mechanic and the motorist. They would effectively remove the mechanic's class and ethnicity from the exchange and circumvent any doubts about his knowledge.[41]

The Ford Motor Company also recognized the scientific connotations of its large diagnostic units and the authority they could bring to the service relationship. In late 1934, with 5,450 Laboratory Test Sets sold to dealers' shops across the United States, Ford emphasized to its dealers' service departments the importance of the scientific image: "The selection of the name for this piece of equipment was very carefully considered, and we are sure that its effectiveness will be enhanced by the identification of it by its real name, i.e., Ford Laboratory Test Set. It is *not* a motor analyzer, or a motor X-ray, or any one of the other trade names that we hear used from time to time."[42] While Shakespeare's Juliet may have believed, "That which we call a rose, by any other name would smell as sweet," the Ford Motor Company was certain that the name of the Laboratory Test Set was integral to its function, integral to the company's campaign to re-create the public image of its dealers' mechanics.

Ford followed the introduction of its Laboratory Test Set with an advertising campaign featuring Ford mechanics as sophisticated technicians who used this scientific instrument to diagnose and detect automotive troubles. No longer the descendants of noble but parochial blacksmiths, dealers' mechanics in Ford's ad campaign became lab-coated technicians pushing Test Sets across clean floors. As recently as 1931, in a widely published advertisement for Ford service, Ford had used the nostalgic image of a blacksmith at his forge as a symbol of trust and tradition. Following the introduction of the Laboratory Test Set in 1934, the blacksmith image disappeared from Ford service ads until 1947, when it reappeared in the form of a cartoon line drawing of a burly, unshaven smith putting the wrong size shoe on a startled horse. The caption read, "Right fit is important in car and truck parts, too!" In the context of Ford's campaign to re-create the image of mechanics as scientific laboratory technicians, the ads discarded the blacksmith as no longer a symbol of respect but now a symbol of backwardness.[43]

Conjuring this new image of mechanics as lab technicians proved easier than sustaining it, as Ford soon discovered that its dealers' mechanics did not actually use the Laboratory Test Set. They were not familiar with diagnosing troubles in the manner dictated by the Laboratory Test Set. Using undercover "service shoppers" to check its usage, the company found that "in entirely too many cases . . . the dealers' man was so unfamiliar with the use of the set that he was reluctant to use it."[44] While the instruments and diagnostic technology it incorporated were not new to the industry, the set relied heavily on abstract numerical readings from electrical gauges and vacuum gauges which did not fit the practices of most mechanics at the time.

Diagnosing and repairing automobiles remained largely a visceral process in the 1920s and 1930s. Mechanics could discover broken external parts, loose wires, fluid leaks, and other such problems by sight. They listened to engines for knocks, looked at bearings for wear, and smelled spark plugs to evaluate combustion. They most often used tools that extended the natural capacities of their physical bodies: wrenches and screwdrivers added leverage, feeler gauges heightened manual sensitivity, and stethoscopes focused auditory sensations. Even when diagnosing electrical ignition system problems, mechanics thought in visual terms such as noting the color of the spark or how large a gap it could or could not jump.

This type of sense-based knowledge was aptly described by philosopher Michael Polanyi as "tacit knowledge," knowledge that is difficult to describe verbally or to communicate in writing. Tacit knowledge defies quantification or mathematical abstraction. It is derived from experience and forms the basis of what we commonly call "skill." Sociologist Douglas Harper has called this aspect of an automobile mechanic's skill "kinesthetic," or bodily, knowledge. It is only one part of a modern mechanic's technological know-how, but through the 1930s it formed the major part of an automobile mechanic's ability.[45]

A mechanic's ears provided perhaps his best diagnostic tool. As early as 1905, E. T. Birdsall, a member of the Automobile Club of America's racing board, suggested that a would-be chauffeur or mechanic should "learn to study his machine by his ears as much as by his hands. By listening to the car and noting the different noises, one may be able to detect immediately any flaw in the perfect running of the mechanism."[46] Throughout the 1910s and 1920s writers tried, mostly in vain, to classify and describe the various sounds that could emanate from a car and what these noises meant. A. J. Brennan divided automotive noises into two groups. Group one noises constituted "the sounds produced by the normal running of the motor; these include the roar at the carburetor, the click of the valves, [the] roar of gears and other rhythmic sounds, which are easily recognized." Group two noises consisted of "unusual noises which may be . . . symptoms of trouble." He further divided these symptomatic noises into five types: knocking, pounding, squeaking, "hissing or puffing," and "numerous puffs or pops." Brennan sought to aid mechanics by describing how various mechanical problems, such as pre-ignition, insufficient lubrication, worn parts, and faulty gaskets, could cause these unusual noises.[47] To aid in sound diagnostics, Brennan and other writers suggested special tools and methods for isolating noises, such as a sounding rod, a stethoscope, or simply cupping one's hands over the ears (see fig. 21).

Using sound as a diagnostic tool was not something practiced only in "hick town garages," as Frank Kettering called the numerous small shops in rural communities.[48] At the 1925 meeting of the Society of Automotive Engineers John C. Talcott of the Pierce-Arrow Motor Car Company, presented a paper on the "Diagnosis of Engine Troubles and Chassis Noises" in which he described the "loud, even hum" of a too-tight engine chain, the "dull, heavy thump [that] indicates excessive play in the main bearings," and "a clicking noise that can often be traced to the oil-pump drive." He described "piston slap," "hill rattle," and "gear whining" as well as numerous other noises and the maladies that caused them. He concluded by saying that "the successful service shop must be able to investigate troubles expeditiously and to apply effective remedies with the least possible delay."[49] The ability to ignore the "normal noises" while discerning the source and significance of symptomatic sounds could be learned, however, only through experience. Until one hears piston slap, no amount of written description can accurately convey what it sounds like. Such was the nature of the visceral knowledge that most mechanics relied upon when diagnosing automotive troubles in the 1930s.

Contrarily, servicing automotive electrical systems presented a significant challenge to many otherwise good mechanics because they could not generally employ the same sense-based, visceral knowledge. A mechanic could certainly feel the "snap" of a spark generated by a high-tension magneto, but he would not want to repeat that experience day in and day out while diagnosing ignition problems. On the other hand, the low-voltage systems used for automotive starting, lighting, and accessory systems operated below the threshold of tactile sensation. A mechanic could not sense variations in the flow of electricity in these circuits; he could not "feel" resistance. Knowing electricity required using voltmeters, ohmmeters, and ammeters. These instruments, rather than extending the physical body, translated electrical properties into numerical representations, which could then be interpreted or mathematically manipulated to diagnose the circuit under study. This fundamental difference between mechanical and electrical knowledge left many mechanics uneasy about diagnosing electrical problems and led to the early specialization of such work in auto electric shops.

Thus, when Ford began shipping its Laboratory Test Sets to dealers in late 1934, experience with its type of analytical diagnostic methods was largely restricted to mechanics with specialized experience in automotive electricity. Most general mechanics likely relied upon visceral diagnostic methods and would have been understandably "reluctant to use it," as Ford's service shopper had discovered. This reluctance forced Ford to undertake a massive training program in order to sustain its efforts to create the mechanics–as–laboratory technicians image.

In April 1936 and again in July of that year the Ford service department informed all branches that "the Laboratory Test Set course is paramount, until all dealers mechanics [*sic*], service managers, floor men, and shop foremen have been trained."[50] The Test Set school averaged about sixteen hours of instruction, after which mechanics had to pass a Test Set Examination that required them to draw schematics of simple circuits, interpret meter readings, diagnose what mechanical conditions such readings might indicate, and show they understood how incorrect connections could damage the Test Set.[51] Mechanics achieving high scores on the exam and demonstrating proficiency in use of the Test Set received Laboratory Test Set Certificates. Ford suggested that "such men be given more credit and encouraged further by increased remuneration for their efforts in acquiring such skill."[52] Over the course of 1936, 22,616 dealer mechanics, or 82 percent of all mechanics in the branch territories, as well as 1,310 neighborhood station and associate dealer mechanics passed the Ford Laboratory Test Set examination.

Ford's sixteen-hour course, however, did not create tens of thousands of laboratory technicians or automotive electricians. Only 8,252 mechanics received the Laboratory Test Set Operator certificates indicating proficiency and warranting higher pay.[53] Ford's Test Set school merely trained dealers' mechanics in the basic diagnostic methods that they needed in order to sell more service with the Laboratory Test Set (and not damage the expensive unit in the process). Evidence is thus far lacking to determine how purchasers of Weidenhoff, Stromberg, or Sun diagnostic units incorporated these equally analytical instruments into their diagnostic procedures, but it is doubtful they fared much better than Ford. In trying to re-create the public image of its dealer mechanics as lab-coated technicians, Ford contradicted not only the predominance of visceral knowledge among mechanics but also worked against a rapidly growing public vocational education system that slighted analytical training in math and science in favor of hands-on work in auto shop.

The various methods by which industry reformers attempted to deal with the service problem in the interwar years—licensing, redesigning, rationalizing, and "scientizing"—all amounted to ad hoc tinkering with the sociotechnical ensemble emerging around auto repair. The various solutions pursued highlight and emphasize the extensive linkages of technological ensembles. The historical actors involved turned to the various social tools at their disposal: state and governmental apparatuses, scientific management, engineering design, and the white lab coat of the scientist. Still, none of this tinkering proved sufficient to do what advocates intended. Substantial and stable social structures had grown up around repairing the now forty-year-old technology of automobiles—social structures that had devalued automotive repair knowledge and confined that knowledge to a narrow demographic band. This ongoing legacy remained at the core of the service problem, and none of the attempted reforms of the interwar period changed that social and technological reality.

CHAPTER 6

Suburban Paradox

Maintaining Automobility in the Postwar Decades

"Why, if my kid wanted to be an auto mechanic, I'd take my hammer and knock him in the head," groused a middle-aged Iowa mechanic at the close of the 1960s.[1] In the quarter-century following World War II the imagined freedom and prosperity of suburban automobility ran hard up against the social hierarchy of technological knowledge. World War II and the years that followed established a milestone in the development of America's automobile culture and suggested a bright future for all things automotive. On the battlefield a flood of military technology flowed from Detroit's converted automobile factories into Europe, the Pacific, North Africa, and even the Soviet Union, eventually proving decisive in the global industrial slugfest. Wartime spending ended the Great Depression, and at war's end American consumers engaged in a great car-buying orgy as they were unleashed from rationing and prodded to purchase. Automobile ownership reached unprecedented levels, ushering in an era of automobile enthusiasm, road building, home sales, and increasingly automobile-centered lives. Americans worked, shopped, dined, and vacationed from behind the wheel. Race car drivers and their mechanics achieved celebrity status in a host of new and growing motor sports. The future looked bright for boys who liked to poke around under the hoods of cars.

Yet by the end of the 1960s any such implied promise for the average mechanic remained unfulfilled. In late 1968 the United States Senate launched an extensive, two-year investigation of fraud in the auto repair industry; employers decried a critical shortage of mechanics; large numbers of vocational auto shop graduates chose not to enter the trade; and the public reputation of the automobile mechanic reached an all-time low. A central paradox of technology's middle ground grew glaringly apparent: almost any involvement with automobiles—owning, racing, tinkering—bestowed status, except for those who repaired them for a living.

Wartime Mechanics: GI Joe and Mary Jo

World War II differed from World War I in that automotive technology no longer remained a novelty to most Americans. American car manufacturers produced a record 3.7 million passenger cars in 1940, and the number of registered cars reached a peak of 29.5 million in 1941—eight times the level at the start of World War I.[2] In addition, the social structure of the auto mechanic's occupation, and the institutions for perpetuating that structure, stood well established and relatively stable by the outbreak of World War II. As a result, U.S. military brass could tap a deep reserve of mechanical experience that would be crucial on the battlefield. Historian Stephen Ambrose noted how this relatively widespread mechanical ability among American soldiers contributed to their effectiveness on the battlefield vis-à-vis German troops. The American GIs "replaced damaged tank tracks, welded patches on the armor, and repaired engines. Even the tanks damaged beyond repair were dragged back to the maintenance depot by the Americans and stripped for parts. The Germans just left theirs where they were."[3] The army expanded its motor vehicle training center at Camp Holabird, Maryland, and the training there and elsewhere could focus more on training soldiers in the specific maintenance and repair requirements and procedures of the now more specialized machinery of war.[4]

Even so, wartime labor demands soon exceeded the supply of experienced mechanics and drew a number of women into military and commercial repair shops. Paralleling Rosie the Riveter's move into manufacturing plants, Mary Jo the Mechanic crossed the well-established gender barriers to work in the shop during the war emergency. Effective February 1942, the federal government ordered a halt to passenger car production in order to shift the materials and production facilities of U.S. automakers to war production. This move essentially dried up the supply of new cars for the duration of the war, and therefore Americans had to maintain and repair their existing vehicles more intensively.[5] Also, following the shock of the attack on Pearl Harbor, the rush of men to enlist in the armed forces left a labor vacuum among garages. Additional men left garage jobs to take higher paying and seemingly more urgent war production jobs. Overall, "mechanic manpower" dropped 40 percent between December 1940 and December 1942 according to a survey conducted by the American Automobile Association (AAA). Surveying more than six thousand shops, the AAA reported that garages and service departments turned away 70 percent of the work that came to their doors. The editors of *Motor Service* magazine surmised that "additional men must be trained if this is at all possible, to alleviate the situation."[6] Yet men were scarce, and employers soon turned to women. By 1943 the Studebaker Corporation declared that "women can and must be employed for automotive maintenance service!"[7] Like Rosie the Riveter, Mary Jo the Mechanic donned coveralls when the need and opportunity arose, whether in shops associated with the U.S. military, in the commercial garage, or at home caring for the family car (see fig. 22). Employers faced with turning away urgent war work and paying customers became much more willing to put women to work in their shops.

Some women may have been eager to explore the mechanical world of the repair shop; others may have been forced by circumstance to take over the family business as a husband, father, or brother went into the service. In the spring of 1943 Evelyn Rand of Bangor, Maine, began her journey across the gendered technology threshold when she answered a Civil Service ad for women to take a mechanical aptitude test. She and six other women from her area passed the test and joined a class of forty women from New Jersey, New York, and Connecticut for an intensive, ten-week auto mechanics training course at the Brighton Trade School near Boston. As a twenty-five-year-old divorced mother of two, Rand found it both difficult to be away from home—she did not return to Bangor on weekends with the others for fear she would not be able to leave her children all over again—and exciting to learn something entirely new. "Much of it was difficult for us," she later recalled. "For instance, when they started explaining carburetion that was quite a thing to be able to understand all of that. But our studying together and our grilling each other with questions helped. . . . As we progressed each day, it all made sense to us, and we became delighted as we mastered the mysteries of mechanics and the workings of a vehicle, and we found we could fix whatever was wrong."[8]

Following their training, Rand and the other Maine women returned to Bangor and worked at the Army Ordnance shop there, at first as "mechanics helpers" but soon upgraded to "mechanics" with pay comparable to the men in the shop. They wore coveralls while working, but as war workers they were entitled to wear a dress uniform after hours, with "a well-fitted little jacket, and a jaunty little cap, and an A-lined skirt [cut] just below our knees." Their sleeve sported a diamond-shaped patch with the letters WOW for Women Ordnance Workers. "You see, the Army had WACS, the Navy the WAVES, the Coast Guard had SPARS, and we were the W.O.W.s," Rand recalled with a laugh. "We did all types of work on ten-wheeler trucks, jeeps, Dodge weapon carriers, and staff cars. We overhauled engines, did tune-ups, relined brakes, greased vehicles, and whatever was written up on the job order by the inspector." The ninety-pound Rand proved adept at this type of work, earning the nickname "Mighty Mite" from her male coworkers and a promotion into the machine shop, where she overhauled carburetors, generators, distributors, starters, and wheel and master cylinders; relined brake shoes; and turned down brake drums. When the Bangor shop closed down in the summer of 1944, Rand's boss gave her a pay raise and a transfer to Dow Air Base to continue rebuilding and repairing ordnance vehicles.[9]

The precise number of women who joined Rand in the auto mechanic's trade during the war is unclear. They were surely numerous because a large number remained in the workforce five years after the war. The U.S. Census Bureau counted 4,082 female auto mechanics in 1950—an increase of more than 340 percent over 1940 figures and more than four times the rate of increase in male mechanics over the same period. Yet this still did not bring their numbers close to 1 percent of all mechanics. The 1950 census recorded the tail end of a wartime spike in female mechanics rather than a long-term trend. Rand, for example, had worked as a mechanic for the Ordnance Department for a little over three and a half years, and at war's end she wanted to remain a mechanic: "Oh, I loved it. I really did. I thought it was the most wonderful thing to be able to do that. I really did. . . . It was fascinating to be able to be able to repair the troubles on a vehicle." "But of course we all knew the war was over. And of course, they had to disband all this. . . . I was heartbroken. . . . I was heartbroken for my own sense of personal loss, but you had to be happy that the whole [war] was over with, too. . . . I wanted to stick with the same work. I was small, but strong. But a *woman;* no way!"[10] The strong gender barrier against female mechanics snapped quickly back into place. By 1960 the U.S. Census detected only 2,305 female auto mechanics, a return to prewar levels as a percentage of the occupation.[11] While Rosie the Riveter may have carved out permanent postwar footholds in some manufacturing industries, Mary Jo the Mechanic secured only a fleeting presence in the repair shop.[12]

For soldiers the U.S. government exhibited more concern with demobilization in 1946 than it had in 1919, but ironically, the government's efforts to help men return to civilian life probably made the labor market even tougher for auto mechanics. The Department of Labor predicted a rapid postwar growth in employer demand for automobile mechanics but warned at the same time of a job market glutted with prospective auto mechanics. With the return of domestic automobile production and the end of rationing, Labor Department analysts expected that automobile sales, registrations, miles driven, and the employment of mechanics would all rise sharply between 1946 and 1950. During the war the number of employed mechanics had dropped from about 377,000 in 1940 to 225,000 in 1945, and their numbers were expected to increase rapidly to 450,000 by 1950.[13]

The War Department estimated that 140,000 of its enlisted men had experience in the auto repair field prior to the war and surmised late in the war that about 200,000 enlisted men and 25,000 civilian employees performed work in the armed services "comparable to the duties of automobile mechanics in civilian life." Furthermore, National Automobile Dealers Association (NADA) wartime surveys indicated that even greater numbers of mechanics left repair shops to work in war production jobs than had gone into the armed forces.[14] With the end of hostilities, then, thousands of men inside and outside the armed services would likely seek employment as auto mechanics in the private sector. "Taking into consideration all these groups of potential entrants into the labor market," reported the Labor Department, "the conclusion is inescapable that job seekers will be more numerous than job openings in the occupation for several years after the war. . . . For [the] less skilled worker and, still more, for persons without any previous experience in the occupation, the Nation-wide employment outlook is unfavorable."[15]

Nonetheless, various agencies within the federal government helped move demobilized soldiers into the occupation. Employers who wanted to hire skilled veteran mechanics could get assistance in identifying qualified former GIs from one of the 644 local Selective Service Boards. Employers looking for less costly labor could also turn to the government for help. The GI Bill of Rights, well known for its home mortgage and college tuition benefits, provided up to fifty dollars for tools and a toolbox for apprentice mechanics and a subsistence of sixty-five to ninety dollars per month to offset low apprentice wages paid by their employers. As their skills improved, their wages would gradually increase and their veterans' subsistence would decrease. This subsidy of low-wage apprentice mechanics no doubt brought even more entrants to the occupation in the postwar years, contributing to the flooded labor market predicted by the Labor Department's report.[16]

Skilled veterans who wanted to open their own shops after the war also got some help from the government. One of the programs developed by the War Department to help smooth soldiers' transition from active duty to civilian life involved educating interested GIs in the principles of operating a small business, thereby encouraging the "many men [who] are seriously considering running a business of their own after this war."[17] Such men, if successful, would create their own jobs rather than compete with laid-off war workers and other returning GIs. Ideally, they would become employers, creating new jobs, and would not likely join or support union activity as they sought to keep costs down. In support of this educational program the War Department contracted with the Bureau of

Foreign and Domestic Commerce to produce and publish a series of small business manuals, initially for use by the armed services and later made available to the public. Significant among the manuals were *Establishing and Operating a Service Station* and *Establishing and Operating an Automobile Repair Business.* These two volumes assumed the reader was "a thoroughly experienced skilled worker" who did not need to learn how to work on cars. Rather, the manuals walked the reader through business considerations such as "Choosing the Location," "Capital and Credit," "Forms of Organization," "The Advertising Budget," "Offering Wage Incentives," "Controlling for Profit," and "Records and Bookkeeping"—essentially subjects usually left out of high school auto shop curricula. The number of GIs who opened garages and service stations after receiving encouragement, instruction, and even financial support from the government warrants further investigation.[18] Taken together with the publicity generated by the Standard Oil Company's promise to set aside a five million–dollar loan fund for veterans who wanted to own their own filling stations,[19] these postwar programs contributed to growth in the auto mechanics occupation which far outpaced the Labor Department's estimates. Government and corporate programs aimed at helping GIs become independent mechanics and gas station operators placed any women with similar needs or interests at a great disadvantage. The hyper-conformity of many Americans to idealized gender norms in the postwar era together with the GI-centered programs helped to solidly re-masculinize the auto repair shop and service station lube bay.

As war production ended, automakers raced to meet the pent-up demand for new cars, and many who wanted to work with cars could sense that America was entering a prosperous golden age of the automobile in which a good mechanic could not lose. Between 1940 and 1950 private and commercial motor vehicle registrations jumped an unprecedented 51 percent, to 48.6 million cars, trucks, and buses.[20] Five years later U.S. automakers shipped nearly 8 million new cars annually—more than twice the prewar level—and new car sales accounted for 20 percent of the nation's gross domestic product. In 1955, according to historian James Flink, "the United States produced about two thirds of the entire world output of motor vehicles," and for twenty years after the war U.S. automakers as a whole reaped profits on their investments nearly twice as generous as the average profits for all American manufacturing firms.[21]

Americans also grew significantly more dependent on their cars in the 1950s and 1960s. Before World War II Americans had certainly embraced automobiles and had begun to consider them essential to their activities. An inspector for the U.S. Department of Agriculture in the 1920s encountered a woman whose rural home had no indoor plumbing and asked her why her family had purchased an automobile rather than installing sanitary plumbing. She replied, "Why, you can't go to town in a bathtub!" She sensed that the automobile could do more for her family than a bathtub could. As more Americans like her made similar decisions, their choices began to alter the physical and social geography of both urban and rural space. In historian Joseph Interrante's words, "What began as a vehicle to freedom soon became a necessity . . . it became a prerequisite for survival."[22] Indeed, through the years of the Great Depression new car sales decreased dramatically, but total automobile registrations remained little changed. Despite economic hard times, Americans chose to cut corners in other areas of their family budgets rather than get rid of their cars.

Following the disruptions of depression and war, Americans experienced not a new dependency on their automobiles but a significantly greater dependence. In the twenty years following 1950, when the number of Americans in the workforce expanded by 42 percent, the number of automobiles registered in the United States jumped by 221 percent.[23] By 1963, 76 percent of all American workers traveled more than a quarter-mile from home to work, and of those, 82 percent relied on their automobiles to get to their jobs and back.[24] If they were not commuting to and from work, Americans increasingly depended on their cars to shop, dine, vacation, and even worship. A suburbanized, auto-dependent America would need an army of mechanics to keep their vehicles running.

Boys under the Hood

For many Americans automotive enthusiasm in the postwar decades meant more than simply increased mobility. They embraced automobile technology for its own sake; they tinkered under the hood; they modified, altered, repaired, raced, and lovingly admired automobiles themselves in all their technical details. This excited under-hood activity introduced many young males to the mechanic's occupation. It also brought new business opportunities, spawned new leisure pursuits, and garnered more public attention for the occupation in general. Women continued to participate at the margins, but they worked against now strong gender stereotypes about cars, car mechanics, and mechanical knowledge. The auto mechanic's occupation figured in an overall automotive enthusiasm that rippled through American society as thousands of Americans, inside and outside the mechanic's occupation, participated in a wide range of under-hood activities (see fig. 23).

Under-hood activity increased, in part, simply because automobile ownership reached unprecedented levels and, more significantly, because the artifact still retained a high degree of postproduction interpretive flexibility—it could mean different things to different users.[25] Detroit's relatively conservative—some would say inefficient—engine designs during the 1940s and 1950s left ample room for user modifications and aftermarket improvements. During the postwar car boom machinist-entrepreneurs such as Jack Engle and Ed Iskenderian experimented with regrinding stock Detroit camshafts to modify valve timing, lift, and duration for increased engine performance in specific situations. Others experimented with modified carburetors and intake manifold configurations, header and exhaust set ups, milled heads, bored cylinders, and oversized pistons. Some toyed with alternative rear end gear ratios, specialized suspension setups, even entire engine and transmission swaps. George Barris and fellow customizers manipulated sheet metal, engines, frames, glass, and upholstery to create one-of-a-kind custom artworks. Modifying stock Detroit iron was not new, but the scale and scope of these activities during the two decades after World War II was. Entirely new sanctioning bodies such as the National Association for Stock Car Auto Racing (NASCAR) and the National Hot Rod Association (NHRA) organized, promoted, and profited in their respective racing niches, and virtually every form of motor vehicle racing enjoyed increased popularity from both spectators and participants. Those who engaged in this surge of automotive enthusiasm poked around under the hood in search of more horsepower, better performance, and, hopefully, bragging rights at the local race track or drive-in soda fountain.[26] Not all of these enthusiasts worked as auto mechanics, and not all auto mechanics participated directly in this flush of postproduction activity. Nonetheless, the activities of racing, modifying, and repairing cars overlapped a great deal and increased the profile, if not the status, of the occupation for a generation after the war.

The career of Smokey Yunick, hailed by *Popular Science* magazine as "America's most famous mechanic," exemplifies this confluence of the mechanic's occupation, the auto racing culture, and popular enthusiasm for under-hood activity. Yunick rose in classic Horatio Alger style from being the poor son of immigrant parents to national prominence in the automotive world, respected by racers, engineers, and everyday boys under the hood. In background and skills he was both typical and exceptional among automobile mechanics.

Named Henry Yunick at birth, he came of age in Depression-era Neshaminy, Pennsylvania, north of Philadelphia, where his Eastern European immigrant parents struggled to make ends meet.[27] He exhibited his mechanical aptitude early in life by converting an old car into a tractor to replace the stubborn chore horse he used to plow the family's land. While not lacking in academic ability, he lost interest in education and dropped out of high school.[28] Yunick's first paying job as a dropout involved digging "a 150-foot tunnel under a cement garage floor for a steam line" at a Ford dealership in nearby Doylestown. Something about this young laborer must have appealed to his employer, for the dealership soon offered to keep him on and train him as a mechanic. His trainee wages were meager, so he took an additional night job in town at Hall's garage. "Hall supervised me enough to get by," Yunick later recalled, "but half the time I was lost." Nonetheless, while working at Hall's, Yunick decided he wanted "to be connected with engines or anything that went fast." He taught himself most of the basic science he had missed in high school by reading physics and chemistry books. He taught himself more about cars by reading all the manuals he could, asking lots of questions of his peers, and by "'cut-and-try' experience." Fleeing minor trouble with the law at home, he worked a short stint as a Chrysler mechanic in Fayetteville, North Carolina, after claiming to know "a little" about the troublesome Chrysler Airflow model. As World War II approached, he returned to Doylestown and worked in a machine shop, roughing out 90 mm antiaircraft gun barrels on a "big son-of-a-bitch" lathe. In final peacetime years he dabbled in motorcycle racing—where his bike's oil-drenched exhaust earned him the nickname "Smokey." Like many young mechanics at the time, then, Smokey moved from the garage to the war production shop, eventually making room for Mary Jo the Mechanic during the war. Once he joined the service, he fixed and flew airplanes for the duration of World War II.[29]

Smokey's postwar experience was also typical of American mechanics. Having married, he bought a 1938 Ford and a house trailer and landed a mechanic's job at a New Jersey Ford dealership. But he and the service manager, Al, did not see eye to eye. As Smokey recalled it, Al displayed a callous disrespect for the mechanic-veteran by repeatedly and needlessly opening the large overhead door near Smokey's work area on bitterly cold winter days. Perhaps Al sensed the eagerness of other young men waiting to fill Yunick's bay and felt no need to cater to the whims of a proud and fussy mechanic. Fed up with his boss's disrespect one day, Yunick dropped the transmission he was installing, gathered his tools, and busted the controls to the overhead door. Without a job and tired of the cold, he immediately moved to Florida, where he found work as a copilot for Eastern Airlines flying a DC-3 at night. Yet "two weeks was more than enough. By then I knew I couldn't work for anybody and would have to start a business. . . . So I started doing the only thing I knew how to do: fixing cars and trucks in trailer parks and gas stations." Obviously, he knew how to farm and how to fly, but he owned neither land nor an airplane. On the other hand, like many young men in 1946, he could begin fixing cars with very little start-up capital. So, Smokey Yunick joined the thousands upon thousands of veterans who turned to the mechanic's occupation after the war. Nationally, the number of employed mechanics reached 650,607 by 1950, a 72.6 percent increase over 1940. Plus, if one recalls that during the war the occupation shrank to about 225,000, then total growth from 1945 to 1950 approached 300 percent.[30] Attractions to the occupation remained similar to those in previous periods—mechanical enthusiasm, ease of entry, and the potential for self-employment—and were now buttressed by Uncle Sam's efforts to help veterans reenter the job market. Smokey does not appear to have sought federal help to open his shop. Rather, after a stint of itinerant repairing, he worked out a shared space arrangement with a blacksmith in Daytona—Smokey helped with a little blacksmithing in return for a place to work on his own customers' cars. In six months time he saved four hundred dollars and bought some riverside swamp land on which to erect a shop. By 1947 " 'Smokey's Automotive' had a roof on, electricity, a phone, and two locking doors." Emblematic of his self-confidence—and perhaps also of his contempt for others less mechanically gifted—the following year he adopted the motto he would make famous among car enthusiasts: "Best Damned Garage in Town."[31]

Smokey's decision to move to Florida, and to settle in Daytona in particular, was fortunate because he now worked and lived in one of the focal points of the growing postwar auto racing world. If he had not gotten involved with Bill France and the birth of NASCAR, it is unclear if Smokey would have been satisfied simply running a small auto repair shop. Following the excitement and danger of flying in World War II, Yunick wondered: " 'How can a 20-year-old man, who has lived a very fast and interesting life, racing motorcycles, working on airplanes, autogyros, fly with French Foreign Legion, the Flying Tigers, B-17s in Africa and Europe, China, Burma, India and the Pacific, and seen so much trouble and dying, or been in so many experiences where it all hung on a thread . . . and lucked out [*sic*]. How can such a man be happy in Florida?' Right in here a racer is born. It was the only thing I could really get a thrill out of. . . . I found what I needed to replace the Air Force excitement."[32] Perhaps this need for excitement to replace the adrenaline of war helps explain the postwar spike in automobile racing generally. For Smokey it would shape the rest of his life. He claimed his wife would not condone his driving in races, so he focused on wrenching for others. He started by helping fellow Daytona racers Marshall Teague and Fireball Roberts and soon built race engines for the Hudson Motor company. By the early 1960s Yunick had gained the respect of racers from Daytona to Indianapolis as well as Detroit engineers and executives, who eagerly sought his opinion. As one journalist put it in 1964, "Cars built or readied by [Smokey Yunick] hold more records than a city-hall filing cabinet." By then his cars had won the Southern 500 at Darlington, the Atlanta 500, the Daytona 500 (setting the record at the time for "the fastest 500 miles ever run anywhere, any time, by any kind of car"), and the Indianapolis 500.[33]

Smokey Yunick was a mechanic's mechanic, a hero mechanic, the idol of countless numbers of boys under the hood. Furthermore, he was a symbol of his time and trade for another reason. Like auto mechanics before and after the war, he relied on a highly developed visceral knowledge of automobiles and engines. Competing engine builder Red Vogt was certain that "Smokey's genius rests in his ears": "There are lots of us who know as much about an internal-combustion engine as Smokey does. . . . It's those damn ears of his that beat us. We do everything to a motor that he does—then he puts his head down and listens to it. What he hears gives him the edge."[34] As another observer put it, "One of the most familiar sights in racing is that of Smokey, half buried under the hood, boots poking out, listening to a racing engine turning over" (see fig. 24). According to Yunick, "Its the only way I can make a final decision. . . . My ears are better than any micrometer I have."[35] Smokey knew cars and engines the way most mechanics at the time knew engines; he just did it better than most.

Thus, in 1959 it made perfect sense for Bob Crosby, an editor for *Popular Science* magazine, to approach Smokey about writing a monthly column to compete with Tom McCahills' auto column in *Mechanix Illustrated.* Comparing the columns of McCahill and Yunick highlights the kind of blue-collar masculinity that helped readers embrace Yunick's mechanical authority and marked him as a "real" mechanic. McCahill often answered letters about light and personal issues, such as the name of his dog, and was typically photographed wearing a sports coat and slacks, even when looking under a hood. He spoke of having owned a garage at one time but was not apparently a practicing auto mechanic. Rather, he was a journalist and professional test driver of new models, about which he rarely wrote anything but praise.[36] When it came to racing, McCahill was a bit of a dandy, setting a speed record for sedans in his large, luxurious Jaguar and then writing about his exploits for *Mechanix Illustrated.* Yunick, if he received personal letters did not bother to answer them in print. He remained a practicing mechanic the entire time he wrote for *Popular Science* and was almost never photographed in clothing other than his trademark workshop shirt, pants, and cowboy hat. The images of Smokey published alongside his column showed him in the environment readers would expect to see a "real mechanic," in grease-smeared pants working on engines. He did not test drive Detroit's new models for the magazine—his reputation for blunt honesty and confrontation would have made automakers quite wary of sending him new models to test. He consulted with Detroit engineers and circulated among NASCAR and United States Auto Club (USAC) officials, but he was never a "suit." Although he had been born in the North, he personified what historian Pete Daniel describes as the "low-down culture" of his adopted South, and his mechanical genius and racing credentials stood unmatched by anyone willing to answer the public's letters regularly, month in and month out.[37]

Smokey eventually agreed to Crosby's proposal and for twenty-eight years served as what he called the magazine's automotive "Dear Abby."[38] Beginning as a regular monthly feature in March 1964, "Say, Smokey—," became a forum for readers' automotive questions and concerns. He answered readers' letters himself, never using a ghostwriter. The published letters and answers, together with other automotive how-to articles that appeared in *Popular Science* and similar magazines, provide a revealing glimpse of what readers of mass audience do-it-yourself magazines were doing under their hoods and what America's most famous mechanic expected they could do.

Evident in the do-it-yourself monthlies, readers' under-hood activity ranged from routine maintenance and repair work to radical modifications. Some clearly used Smokey as a second opinion when they were dissatisfied with their own mechanic (much as callers do today with the weekly *Car Talk* radio program on National Public Radio). More commonly it appears that when readers purchased a new car they looked under the hood with a purpose: to see where the oil filter was and how hard it would be to get at the spark plugs. Articles such as "How to Take Care of a '57 Chevy" provided detailed, illustrated instructions to "Saturday mechanics" on "how to adjust [the] ignition, carburetor, lights, body alignment, [and] brakes" on Detroit's latest models.[39] Other articles provided illustrated instructions on seasonal maintenance routines. Typical was *Popular Science*'s 1963 "Spring Car Care" feature, which walked readers through carburetor adjustments, including fast idle, choke linkage, hot idle, and idle mixture. It also provided a diagram for making and using a timing light and instructions on adjusting ignition points with a feeler gauge.[40] Carl Marten of Parkland, Washington, recognized this interest and ability among his customers when he offered a "self-service" option in his garage in 1951. For fifty cents per hour motorists could rent a space in his garage, consult his service manuals, and use the shop's tools to do their own service and repair. He appears to have been correct in his assessment of his market, for he reported a threefold increase in gross revenues after offering self-service.[41]

Users of Marten's self-service garage could pay extra for his expert consultation on problem jobs. Readers of the do-it-yourself monthlies who lacked the experience of practicing auto mechanics and without easy access to the oral exchange of knowledge common in most repair shops used columns such as "Say, Smokey—" to ask questions about particular problems they encountered.[42] In the first regular "Say, Smokey—" column James W. van Gundy of Rockford, Ohio, sought help with his 1958 Ford V-8: "It spark-knocks [or "pings," indicating preignition] all the time." He had already replaced the distributor, the vacuum advance, and "set the timing in various ways. Nothing helps, even on high-test gas." So, he sought help from Smokey, who instructed him to dig even deeper into his engine: pull the heads and clean the carbon buildup on the pistons and compression chambers. He also advised checking to see that the head had not been milled too much by a previous owner, thus increasing the compression ratio too much for pump gas.[43] We cannot know with any certainty if van Gundy did the distributor and timing work himself or if he later removed the heads as Smokey advised. Yet when he and other readers wrote to Smokey, they frequently referred to working on their cars in the first person, as work that they performed, and Smokey replied to such letters as though they could do their own work.

Other readers turned to Smokey for advice on more extensive modifications they wanted to make on their vehicles. Roger Paulsen, of San Francisco, had decided to mill the heads of his 1961 Impala to boost the compression ratio to 11.1:1 and wanted Smokey's opinion on whether the crankshaft would need "beefing up." John Milks, of Kansas, sought advice on the best camshaft for increased performance in his Pontiac engine, and George Perroni, of Chicago, was "hopping-up" his Ford Galaxie "for performance" and wanted Smokey's advice on installing a limited-slip differential.[44] To all of them Smokey offered encouragement and practical advice.

Radical alterations considered by readers did not always involve speed. Al Sorenson, of Salt Lake City, wanted to put a small V-8 engine into his "underpowered" Ford Econoline van, clearly not a race setup. Jim Nichols, of Bedford, Virginia, wrote Smokey: "I'd like to install a front-seat roll cage and shoulder harnesses in my Volkswagen bus. I don't race, but want it for protection if somebody hits me." Smokey thought this was a good idea and suggested Nichols "head for the nearest shop where they work on racers for NASCAR or USAC." David Hogan, of Binghamton, New York, had a problem more likely encountered by other readers. The brakes on his Ford F250 pickup truck were "fine for light loads and speeds under 50 m.p.h., but pedal pressure goes sky high with a loaded camper at 65." Smokey's reply: "Hope you haven't committed suicide with that arrangement before you read this! Better get a power cylinder installed, and change to a premium lining such as Raybestos or Grey Rock. It holds up even on race cars."[45] With the popularity of automobile vacationing in the postwar decades, many motorists sought help from mechanics and aftermarket suppliers to boost their vehicle's towing capacity by installing tow bars, load leveling shock absorbers, bigger radiators, bigger alternators, dual exhaust, and as David Hogan discovered, better brakes. Smokey's advice throughout the 1960s and 1970s was always pragmatic; he assumed the reader understood the basics of automotive maintenance and repair. His column, as well as much of the do-it-yourself literature and many of the aftermarket suppliers, expected that a significant number of readers/motorists could do their own work or could speak knowledgably with a mechanic who did the work for them.

A Wrenching Paradox

Although Smokey Yunick may have risen in Horatio Alger fashion from poor son of immigrant parents to "America's most famous mechanic," his celebrity status among enthusiasts did not transfer generally to the auto mechanic's occupation. Readers of do-it-yourself monthlies may have respected Smokey's masculine mechanical knowledge and no-nonsense style, just as they celebrated the televised skill and speed of famous racing pit crews such as the Wood brothers at Daytona or Dan Gurney's crew at Indianapolis.[46] One study estimated that the volume of do-it-yourself auto repairs increased from 15 percent of all auto repairs in 1959 to 24 percent by 1965.[47] Nevertheless, the respect and the mutual understanding between motorists and mechanics, which might be expected from the overall increase in under-hood activity in the postwar decades, did not find its way into the average repair shop.

Thousands of mechanics entered technology's middle ground in the postwar decades, and like Smokey, they tried to answer and correct motorists' mechanical problems day in and day out, but as the 1950s turned into the 1960s and 1970s, they discovered that the occupation gradually grew more complicated and less rewarding. After the initial postwar rush into the trade, growth in mechanic employment slowed to 5.2 percent between 1950 and 1960 then picked up to 29 percent from 1960 to 1970. Automobile registrations continued to grow, however, at 50 percent and 46 percent in those respective decades. The divergent growth rates between mechanics and vehicles meant an increasing ratio of cars to mechanics. In 1930 that ratio stood at about 72 cars to 1 mechanic. In 1940 it nudged up to 85 to 1. The rush of postwar mechanics brought the ratio back down to 75 to 1, only to return to a steady rise thereafter, reaching 107 to 1 in 1960, 120 to 1 in 1970 and 194 to 1 in 1990.

On the surface a rising ratio of cars to mechanics might mean either cars were getting better and needed less time in the shop or mechanics were entering a period of increased demand for their labor in which their wages and status might be expected to improve. Neither held true. Certainly, advances in automobile engineering and manufacturing reduced or eliminated some maintenance operations. The introduction of hydraulic valve lifters in the late 1940s, for example, virtually eliminated the periodic valve adjustment. Improved cylinder valves and seats and the better flow of coolant in the block and heads greatly reduced the need to regrind valves. Other changes, however, mitigated such advances in reliability.

Industry insiders correctly viewed the increasing ratio of cars to mechanics as evidence that the service sector was not keeping pace with the proliferation of automobility. Rather than becoming simpler, maintenance and repair was becoming more complex as the cars coming out of Detroit grew more diverse. The history of the Model T revealed the dire consequences of mass production without product differentiation as Ford sales plummeted in the late 1920s. With no major stylistic changes in nearly two decades, the Model T became the plain oatmeal of automobiles, sensible and inexpensive but no longer sought after.

Postwar Detroit therefore embraced and extended Alfred Sloan's model differentiation strategy based on price, style, and accessories, initiated at General Motors in the late 1920s. Sloan combined this strategy with the planned obsolescence of all models through annual model changes. While Ford saturated the market in the 1920s by driving down production costs and the sale price of the Model T, Sloan sought to instill in the customer a desire to purchase a new car before the old one had worn out. Using this strategy, GM dislodged Ford from sales leadership in the 1930s, and the momentum of the automotive marketplace shifted from Fordism to Sloanism by the outbreak of World War II. Economic depression and war dampened its full implementation, but postwar affluence provided the perfect economic environment for the full implementation of Sloanism, and the remaining American automakers got on board. As one Ford executive described his styling department's objective, "We design a car to make a man unhappy with his 1957 Ford 'long about the end of 1958."[48] In this environment automakers and their dealers aggressively marketed highly profitable accessories that brought new service and repair demands. By the 1965 model year 81 percent of American cars rolled off the assembly line equipped with automatic transmissions, 74 percent had radios, and 60 percent had power steering. By the end of the 1960s half of all new cars also had power brakes and air-conditioning.

These enhancements and improvements appealed to motorists generally but, together with the annual model changes, complicated service and repair. Clutch replacement jobs became much less common, but mechanics had to learn to diagnose and repair automatic transmissions in which hydraulic pressure transmitted the engine's power, not a conventional clutch and gearbox.

More cars in more models with more options and accessories meant more service, not less. Mechanics, however, did not generally benefit from this increased demand as might have been expected. No doubt many more had jobs than before the war, and they were not sitting on the workbench for week-long stretches with no customers, as had often been the case during the Great Depression.[49] Many no doubt liked working on cars and may have enjoyed witnessing the technology and culture of the postwar auto boom from their unique vantage point under the hood. Yet their wages and their status did not mirror America's broader auto enthusiasm and economic growth. Instead, the auto repair industry experienced further segmentation and increased competition as automakers experimented with new car warranties and as mass merchandisers and franchised specialty shops sought to profit from maintaining suburban automobility.

As before the war, the texture of the auto repair industry in the postwar decades initially featured a few large urban shops and dealership service departments with divisional organization employing a crew of mostly white male mechanics. Far more numerous, however, were the small, independent repair shops employing just a few general mechanics. In 1962 these small shops with one to three employees accounted for 70 percent of all automobile repair shops.[50] The workforce remained dispersed and unorganized, exercising no effective control over entry into the occupation and no licensing or certification of competence and still receiving a relatively narrow band of recruits reflecting social and cultural expectations created in the first half of the twentieth century.

African Americans continued to struggle with discrimination in some white shops, while others maintained a separate black automobile economy in communities across the United States.[51] One black business publication celebrated Wiley McRae, taxi company entrepreneur in 1946, as "a solid example of race progress." With fifteen cabs, thirty-two drivers, and two mechanics among the forty-one employees, McRae's company—or ones such as Crosby's Taxi in Wilmington and Thurman's Garage in Hallsboro—provided employment opportunities for black mechanics in North Carolina.[52] Julius A. Hunter, in Whitakers, North Carolina, opened his own repair shop in the mid-1930s with the financial backing of a white doctor who had faith in Hunter's mechanical skills. By 1948 Hunter employed four black mechanics in addition to himself, and by avoiding politics and ignoring "the rantings of professional Southerners of the 'white supremacy' type," he built a business that attracted both white and black customers.[53]

Adding to this potpourri of auto repair firms in the late 1950s and early 1960s were the mass merchandisers, such as Sears, Roebuck, and franchised specialty shops, such as AAMCO transmissions and Midas Mufflers. Sensing the needs of suburban motorists, Sears, J. C. Penney's, Montgomery Ward, and other large department stores began offering basic automobile maintenance and repair services at their own branded auto centers.[54] Often located in purpose-built structures in shopping mall parking lots adjacent to their anchor stores, these auto centers hoped to appeal to motorists' convenience. With 76 percent of U.S. households owning automobiles, 13 percent owning two or more cars, and with women accounting for 39 percent of all drivers in 1961, mass merchandisers tried to make it easy for suburban families to get their cars serviced.[55] Customers could drop their car off for service while they shopped or dined at the mall, or they could wait in the clean, air-conditioned customer lounge, supplied with popular magazines and eventually television sets to occupy their time. Like larger dealerships, these auto centers employed clean-shaven service writers to mediate between the customer and the mechanic in hopes of both easing tensions and facilitating more service sales.

Mass merchandisers introduced considerable competitive pressure in the auto repair industry. They could tie auto service advertising into already well-funded local, regional, and national advertising campaigns. They used their purchasing power to gain discounts on tires, batteries, and accessories—the "easy money" of auto repair—thereby gaining price and profit advantages over small independent shops. They also siphoned off the simple and profitable maintenance and service operations such as oil changes, tune-ups, and brake jobs while referring the difficult, time-consuming, and risky work of major repairs to independent and dealer shops. William Winpisinger, speaking on behalf of unionized mechanics, worried about the competitive pressure auto centers introduced: "They can draw the line on what they can do and not do and . . . they really in essence take the cream off the top of the business."[56] By offering less than full service, retailers' auto centers could also employ younger, less skilled mechanics and pay them lower wages. A GM dealer from Clinton, Iowa, complained that these shops bolstered their profits by using "semiskilled and unskilled help."[57] This practice contributed to the slow growth of wages in the industry relative to demand, a problem that remained serious and unremedied for mechanics for much of the remainder of the century.

Other pressures also kept auto mechanics' pay from keeping pace with other skilled trades. From 1961 to 1967 wages for auto mechanics increased at 3.6 percent per year. Yet in 1970 one Department of Labor official noted that at $3.58 per hour the mechanic's pay remained 75 cents an hour lower than laborers and helpers on an organized construction job. "In fact," he continued, "the average mechanic earns 40 cents an hour less than the unskilled worker on the assembly line in a Detroit auto plant who simply bolts the bumpers on a 1970 Ford."[58] Weak unionization relative to other skilled trades certainly contributed to low mechanics' wages. Organized labor proved no more successful at gaining a foothold in the industry after World War II than it had been before the war. By the late 1960s approximately 80 percent of all mechanics worked in nonunion shops under nonunion conditions. The largest union of auto mechanics remained the International Association of Machinists (IAM), which represented by its own estimate only 10 to 14 percent of the industry, with members concentrated in larger dealership, fleet, and municipal shops in limited geographic pockets such as the San Francisco Bay area, the greater St. Louis area, Chicago, and Cleveland.[59]

The flat rate and piecework pay system presented another obstacle to wage growth. Introduced by Ford in the 1910s and challenged by mechanics in the 1920s and 1930s, it remained surprisingly influential in the postwar period. As a result, auto repair remained the only occupation in which a significant number of skilled workers labored either under a piece rate system or under its influence on prevailing wages. The National Automobile Dealers Association reported that in 1965, 44 percent of American dealerships paid their mechanics by flat rate, but its use varied in proportion to the volume of new car sales. Only 21 percent of the smaller dealers (those who sold between 1 and 99 new cars in 1965) used the flat rate system, while a little over a third paid an hourly rate and a bit less than a third paid straight salary. Ten percent of the small shops paid salary plus commission (either on parts or labor or on both). Of the mid-sized dealers with sales between 300 and 399 for the year, 64 percent used the flat rate system, 14 percent paid hourly, 12 percent straight salary, 9 percent combined salary and commission, and 1 percent paid straight commission. Large-volume dealers selling 750 to 999 cars per year paid flat rate 79 percent of the time, hourly 17 percent, straight salary 3 percent, and 1 percent by "other" means.[60] Similarly detailed numbers are lacking for the independent repair shops, but qualitative sources indicate that the larger shops were more likely to tie their payrolls to the flat rate system, while the numerous small shops were more likely to use hourly wage, salary, or some combination.[61]

For many of the same reasons that it had failed to fix the service problem in the 1920s and 1930s, the flat rate system created problems in the 1950s and 1960s.

It encouraged quick and incomplete work. It discouraged diagnosis in favor of doing only what the work order authorized. It penalized experienced mechanics for taking the time to guide younger mechanics. It encouraged the replacement rather than the repair or rebuilding of parts. More important, it kept mechanics' wages down relative to the demand for their labor. In 1970 the Federal Trade Commission referred to the flat rate / piecework system as "the almost unbelievable method dealers employ in compensating mechanics. . . . The marvel of this system is its ability to survive. . . . Its only possible justification is as an incentive device to get work done quickly. But there it is being substituted for proper management controls."[62] Nevertheless, the National Automobile Dealers Association defended the flat rate system, declaring that "flat rate manuals are of genuine aid to the customer" because they allow set-price estimating of repair work before the work begins.[63] Mechanics expressed their displeasure with factory-determined times for certain operations they felt were set too low. Yet a NADA-sponsored study in 1968 claimed that mechanics beat the factory flat rate time on 74 percent of the jobs and that "the average mechanic books approximately 9.4 hours for each 8.0 hours he works."[64] Union leaders condemned the system as exploitative and archaic but admitted that "the majority of mechanics not only accept it but would very probably oppose elimination. Their incomes have been geared to flat rate schedules for so long that they would rather argue about whether any given flat rate is adequate than fight to replace [the whole system] with higher hourly wages"[65] So, mechanics individually liked flat rates as long as the mix of beatable and unbeatable flat rate jobs balanced in their favor. Still, service managers could manipulate the assignment of jobs to mete out shop justice or favoritism. Moreover, when mass merchandiser auto centers began siphoning the cream off the top of the business in the late 1950s and early 1960s, the union claimed that mechanics' income based on piecework declined.

Finally, as American automakers embarked on a "warranty race" in the early 1960s, warranty repair started flooding the dealerships, and warranty work was notorious for paying mechanics poorly. Before World War II automakers offered no more than ninety day or 4,000-mile limited warranties, but beginning in 1961 all four major U.S. automakers increased their new car warranty to twelve months or 12,000 miles. Then Chrysler upped the ante to a five-year or 50,000 mile "power train warranty" on its 1963 models. Over the next few years each automaker tweaked its coverage in an attempt to gain market share until they all offered five-year or 50,00 mile warranties on their 1967 models.[66] Engineering and production quality did not justify increasing warranty periods much beyond the twelve-month or 12,000 mile mark. The increases were purely marketing driven, and the dealership service departments received the brunt of the public's scorn when it became apparent that the cars were not defect free, as implied by such warranties. One NADA representative described the extended warranties as a failure and "the source of cost, conflict, and confusion to the dealer, manufacturer, and customer."[67]

To mechanics extended warranties meant increased demand for their labor but at a reduced rate of pay because manufacturers only reimbursed dealerships for warranty service based on strict, factory-derived flat rate times and a discounted labor rate. Dealers by and large then passed that loss down the line to their mechanics in the form of lower pay for warranty work. Flat rate warranty work hindered wage growth generally because the manufacturer would only pay a set amount for warranty work, and dealers and mechanics had little room to bargain. Thus, if a single dealer's labor rate much outpaced the warranty payment, the dealer lost money. If dealers passed on too great a difference to their employees, they would generate unwanted tensions. It was no small coincidence that most decided it was best simply to stay put at a labor rate close to the low warranty rate. This strategy had a stabilizing influence on pay rates in the industry generally.

All these pressures in the industry resulted in a sinking morale among mechanics. In 1965 Philco, a subsidiary of the Ford Motor Company, produced a scathing report that laid bare the low spirits among Ford dealer mechanics, a malaise many sensed in the industry as a whole. Philco researchers visited the Ford Central Office, the Lincoln-Mercury Division, all the training centers, and "Ford and Lincoln-Mercury dealers throughout the United States" in an attempt to evaluate the company's overall service training system. The report's anonymous authors also reviewed "the general behavior of the Dealer, the Service Manager, and the Mechanic." They found the dealership owners insincere about supporting service training and more concerned with sales and training salesmen than with mechanics' needs. Dealership service managers seemed genuinely troubled by the "mechanic's plight and training needs" but unable to make a significant contribution to improving their situation. The researchers found that "the average mechanic feels that Ford Motor Company hurts his pocketbook with warranty time and wage":

> He feels that the dealer is depriving him of benefits, such as: retirement, paid holidays, profit sharing, and guaranteed minimum wage. The average mechanic feels sorry for himself, sees no opportunity for advancement, has little pride in his work or vocations, and is envious of others in comparable vocations. With little

> or no goals to reach for—he does not show initiative, does not recommend that youngsters become mechanics, and is usually sorry for ever having become a mechanic in the first place. He makes known his desire for training; but is not willing to sacrifice time or money for it. Of the mechanics interviewed, 48% did not graduate high school, 9.3% received vocational school training, and less than 1% had any college education. Lacking in specific job goals, money is the mechanic's major incentive.[68]

This is not the kind of job description that would inspire parents or high school career counselors to send kids into the occupation. Readers of the report must bear in mind that Philco was interested in promoting more training involvement from the Ford Motor Company, which Philco's education division would be in a position to supply. Nevertheless, based on what we know about the developments and pressures within the automobile repair industry in the postwar decades, the "mechanic's plight" description applied beyond Ford and likely beyond dealership mechanics—especially in light of one independent garage owner's comment that "perhaps the one remaining edge they [dealers] have over us is that they are able to attract and hold more mechanics, because of their better hours and more glamorous surroundings in which to work."

Responding to the dissatisfaction expressed by mechanics, Ford, GM, Chrysler, and American Motors all trumpeted new efforts to train their mechanics better and to recruit new mechanics to the trade. Each had limited success but generated many inches of good press. Not surprisingly, the only new nationwide program in the postwar decades aimed at bringing new mechanics into the shop reflected and reinforced the image problem that was dogging the occupation. Under the Manpower Development and Training Act (MDTA) of 1962 the Departments of Labor and Health Education and Welfare coordinated training and provided funding to move the "hardcore unemployed" into the labor market. Vetoed by President Eisenhower in an earlier form, the MDTA became part of the Kennedy and Johnson administrations' war on poverty.[69] Having as its goal the alleviation of poverty and unemployment, the Department of Labor's Manpower Administration Order No. 2-68 defined the program's target beneficiary as "a member of a poor family, and unemployed or underemployed, and is either a school dropout (not a high school graduate), a member of a minority group, under 22 or over 45 years of age, or handicapped in the sense of having a 'physical, mental or emotional impairment or chronic condition which could limit work activities.'" It is doubtful any of the program's sponsors had the needs of the auto repair industry in mind when they drafted the legislation. Nonetheless, by June

1963 Manpower programs trained over 4,000 auto mechanics, and by the end of fiscal year 1969, 63,000 enrollees had been trained as entry-level mechanics.[70]

The Manpower program garnered mixed reactions within the industry. Winpisinger of the International Association of Machinists generally supported it. A high school dropout himself, he believed recruits should be given the opportunity to prove themselves. The IAM coordinated with Manpower to create a nationwide pool of mechanic recruits for the trucking industry, but Winpisinger admitted that the IAM was having difficulty filling the 1,050 entry-level mechanic vacancies it had identified in the industry from the pool of available Manpower recruits.[71] W. Athell Yon, owner of a Charleston, South Carolina, repair shop with fifteen employees, did not support the program. Speaking before Congress on behalf of the Independent Garage Owners of America (IGOA), he criticized Manpower because he and the IGOA did not think their industry needed yet more high school dropouts trained as auto mechanics. Higher-caliber mechanics would be hard to recruit from the ranks of the hardcore unemployed. "This in itself," Yon believed, "tends to downgrade the automotive repair industry."[72] Mel Turner of the Automotive Service Industries Association, clearly had the Manpower program in mind when he criticized "government-funded training programs that become, not business-like programs, but social programs."[73] Given the heightened attention that civil rights issues had gained in the broader political culture of the time, one can only wonder if others in the industry were also concerned about the "minority group" clause in the beneficiary definition and what that might mean for the future status of the occupation.[74]

The new pressures developing in the industry and what the Philco study described as the "mechanic's plight" mitigated any positive bounce the occupation's image may have gained from the fame of Smokey Yunick or other racing mechanics or from the proliferation of tinkering and hot-rodding in the postwar decades (see fig. 25). Whereas the clamoring of eager students in the 1920s had contributed to the establishment of high school auto shop programs, by the 1960s industry observers noted a growing gap between labor demands and the number of new entrants to the occupation. One Department of Labor official surmised that low wages might explain why, out of 98,000 graduates of vocational education auto shop programs surveyed, "only 17,000 actually entered the trade."[75] He was half-right; wages were a big part of the problem. But so was the social identity or image of the mechanic. Many others recognized the difficult chicken-and-egg relationship between image and wages. The editor of *Motor Age* acknowledged: "We are now knee deep in a service economy. [Yet] somehow the service industries in general have not been glamorized to the extent that young men seek careers in keeping things running. . . . Like it or not the skilled mechanic has lost status in our society."[76] A few years later C. D. Crill of the California Bureau of Automotive Repair concurred, "You don't find too many desirable young people that want to become mechanics."[77] Myron Appel, an experienced mechanic who became division chair of Vocational and Technical Education at Cypress Community College in southern California, cited "dismally low and unattractive" wages and benefits as well as the equally important factors of social identity and image, particularly a 58 percent high school dropout rate among practicing mechanics and the public's "lack of respect towards the image or the prestige of the mechanic." As a result, Appel observed, "even the parents limit the number of young people entering our industry."[78]

Plenty of boys enjoyed poking around under the hood in the 1960s and 1970s, but not many aimed for a career in the repair shop. As the National Committee on the Employment of Youth reported in 1964, "There is a certain status in being knowledgeable enough to repair one's own car; there is much less status in doing this type of work for a living."[79] Mel Turner confirmed this assessment in 1971 when he asked a conference room full of fellow industry insiders: "Is there a man in this room who would recommend that his son go into the garage business? Is there really? I mean really and truly. Not somebody else's son, your son."[80] A paradox had emerged in postwar suburban automobility: Americans celebrated their cars—and grew increasingly dependent upon them—but they did not want their kids to grow up to be mechanics.

CHAPTER 7

"Check Engine"

Technology of Distrust

Theodore Jackson entered the office at his boss's request to answer the complaints of a disgruntled customer. Jackson, a Washington, D.C., auto mechanic in the early 1960s, exchanged heated words with the customer, and before long the irate customer hauled off and slugged Jackson right on the jaw, breaking it completely through in two places and landing Jackson in the hospital for eight days. Doctors wired Jackson's jaw shut for six weeks, and he could not work for two months.[1] While mechanics grew increasingly dissatisfied with the conditions of their occupation, their customers grew impatient with the powerlessness they felt in the repair shop. The violence unleashed on Jackson that day symbolized a new level of anxiety, a new pitch of customer frustration with the long-standing "auto repair problem."

Motorists expressed this anxiety in a public sphere increasingly influenced by the consumer rights and environmental movements. Motorists joined environmentalists and consumer advocates in turning the power of expanding state and federal governments toward their concerns in the 1960s and 1970s. Although none could see it in the early 1960s, the confluence of these social and political developments, together with the nascent computer revolution, produced automotive technology in the 1980s which manifested motorists' anxiety. Their increased auto dependency, their continuing distrust of mechanics, and their government's growing regulatory power combined to alter the direction of automotive development toward systems that monitored the behavior of mechanics, motorists, and manufacturers and resulted in computerized engine management systems that brought new repair epistemologies, new ways of knowing the car and its subsystems. By the 1980s these socially and politically determined changes in motor vehicle design fundamentally challenged the social structure that had defined the auto mechanic's occupation for much of the century.

Motorists as Consumers

Fortunately for Jackson's fellow mechanics, most motorists respected the bounds of civil society and did not assault their mechanics—much as they may have been tempted. Bill Matters, a retired real estate agent living in southern California, grew frustrated after taking his new car back to the shop for the same repair three times. On the third visit to the dealership's service department, the 6'5", 280-pound Matters got out of his car and blustered, "Do I have to whip someone to get my car fixed?" Yet rather than follow through on his threat, Matters took his complaint before a California State Senate subcommittee considering, for the first time in decades, legislation requiring certification of automobile mechanics.[2] When mechanical failures hampered the day-to-day activities of motorists, or worse threatened their safety, customers such as Matters demanded action and redress from their elected officials. Politicians and government agents in turn became increasingly responsive to motorists as consumers.

Only slightly older than the automobile, the consumer rights movement evolved in three phases, or "waves," extending from the late nineteenth century to the late 1970s. During the movement's first phase the automobile, seen largely as a luxury beyond the reach of most consumers, generated little concern. Turn-of-the-century progressive reformers provided much of the spirit of the first phase by securing minimum government protections against some adulterated consumer goods with passage of the Pure Food and Drug Act and the Meat Inspection Act in 1906. To the degree that motorists in these years banded together against abuse in the maintenance and repair of their vehicles, they did so as employers of chauffeurs, not primarily as consumers.[3]

Nonetheless, an automobile accident on 25 July 1911 resulted in a landmark product liability case that laid the groundwork for later consumer protection cases. Donald MacPherson and two passengers left Galway, New York, heading toward Saratoga Springs in his 1910 Buick Model 10 Runabout. Just outside Saratoga Springs, traveling at about eight to fifteen miles per hour, the car's rear wheel hit a rut in the road, causing the car to swerve. As MacPherson corrected the steering, defective wooden spokes in the left rear wheel collapsed and sent the car into an uncontrolled spin. The resulting accident threw MacPherson from the moving car and pinned him under the rear axle.[4]

In the ensuing legal battle Buick denied responsibility for MacPherson's injuries, arguing that it had purchased the wheels from another company and that it had sold the car to a dealer in Schenectady, not to MacPherson, and therefore had no contractual obligation to MacPherson. After initially winning the case, Buick lost a series of appeals that culminated in Justice Benjamin Cardozo of the New York Court of Appeals ruling that an automobile becomes "a thing of danger" if negligently made. Thus, he claimed, the manufacturer "is under a duty to make it carefully." The principle of "buyer beware" proved a weaker corporate defense in an age of increasingly complex consumer technology. "In the old days," wrote Justice John M. Kellogg in an earlier appeal of the case, "a farmer who desired to have wheels made for an ox-cart would be apt to inspect the timber before it was painted . . . in order that he might know what he was buying." Yet given "the needs of life in a developing civilization," Cardozo later clarified, "precedents drawn from the days of travel by stage coach do not fit the conditions of travel to-day." *MacPherson v. Buick* extended the principle of legal liability to new circumstances and set an important precedent for consumer rights which would be picked up with vigor in the 1950s and 1960s.[5]

The Progressive Era set in motion another subtle change that would have growing influence on the consumer rights movement and eventually the auto repair industry. The Smith-Hughes Act of 1917—which had established federal funding for vocational education and helped institutionalize high school auto shop—also reflected the home economist movement of the period and provided funding for home economics courses in public schools. In addition to funding the spread of high school courses such as Marriage and the Home and Child Psychology, home economics courses taught generations of young women about "health and hygiene," "clothing problems," "food selection," and "budgeting the income."[6] These types of courses helped sensitize ensuing generations of women to some of the basic tenets of consumer activism: that rational home management and purchasing were duties to be undertaken with deliberation and a concern for obtaining the most value possible in each exchange. Enrollment in home economics courses grew from 31,000 in 1918 to 175,000 in 1930.[7] As they had done in response to growing numbers of auto shop programs, teachers' colleges around the country began offering courses and programs for home economics teachers, creating academic settings for professional home economists and consumer science researchers. In a bit of historical irony, the sex-segregated public education system, which had barred young women from auto shop courses, directed girls into home economics courses that would prepare them to become increasingly influential in the repair shop once the consumer rights movement turned its attention to the automobile.

The economic stresses of the Great Depression ushered in the second wave of the consumer rights movement as the problem of under-consumption took center stage. Franklin D. Roosevelt and many in his administration viewed the loosely defined American consumer as an important, permanent constituent whose interests and problems were to be accepted as the responsibility of government and balanced against the power and interests of industry.[8] The 1930s also saw the growth of important nongovernmental, grassroots consumer organizations. The best-selling book *Your Money's Worth* (1927) by Stuart Chase and Frederick Schlink had criticized deceptive advertising and publicized the concept of neighborhood consumers' clubs aimed at cutting through the puffery of claims and laying bare the facts about products for club members. Schlink received so many inquiries about his own White Plains, New York, club that he eventually "transformed the neighborhood club into Consumers' Research," a nonprofit consumer watchdog organization.[9]

The automobile began to factor directly into the consumer rights movement when Consumers' Research's offshoot and rival, Consumers Union (CU), began testing automobiles in 1936. Like Consumers' Research, CU sought to provide unbiased information to consumers in order to give them more power in the marketplace. As CU cofounder Colston Warne put it, the organization strove "to distinguish truth from fallacy in advertising claims and provide a basis for rational consumer choice."[10] CU initially linked consumer and labor issues, but after being accused by Congressman Martin Dies's House Un-American Activities Committee in the late 1930s of being a "red transmission belt" of communist ideas, CU gradually disengaged from the labor agenda and embraced objective product testing.[11] While it was not removed from the committee's list of subversive organizations until 1954, *Consumer Reports* nonetheless increased its circulation from 50,000 in 1944 to 500,000 in 1950 and 1 million in 1961.[12] By then the union divided its technical division into textiles, chemistry, foods, electronics, automobiles, and special projects. Laurence E. Crooks began testing cars for CU in 1936, initiating one of the key enduring features of the publication. Warne recalled of the years of rapid growth in the late 1940s and early 1950s, "People were buying automobile and household equipment and wanted advice."[13] *Consumer Reports* and rival publication, *Consumers' Research Bulletin,* institutionalized a "distrust of advertising" at the same time that they promoted trust in their own publications by providing "scientific, objective, and impartial evaluations" of consumer goods.[14]

The market for consumer advice and consumer exposé literature grew during the 1930s, while motorists' unease in the repair shop continued.[15] Both came together forcefully in a 1941 article in *Reader's Digest* magazine entitled "The Repairman Will Gyp You if You Don't Watch Out" by Roger William Riis.[16] In a classic case of investigative journalism John Patric and a female assistant drove nineteen thousand miles over a period of five months, visited 347 service stations and garages across the United States, and presented the same car with the same simple problem: a twelve-cylinder Lincoln Zephyr with one of its two coil wires removed. They discovered that three out of five times they were "ripped off" and charged for unnecessary repair work. The article proved quite a success. *Reader's Digest* distributed millions of reprints of the article series, and excerpts and commentary about it appeared in both the popular and automotive trade presses.[17] According to Riis and Patric, parts of the series were even performed as radio dramatizations "over national broadcasting networks in the United States and Canada."[18] By 1942 the authors compiled the series of *Reader's Digest* repair investigations they had done on automobile, radio, and watch repair, along with readers' feedback and illustrations, into a book entitled *The Repairman Will Gyp You if You Don't Watch Out* and a 1949 second edition with the simplified title, *Repair Men May Gyp You*. The success of Riis and Patric's publications reveal that their readers were not just looking for "advice" and information in purchasing new cars such as that offered by *Consumer Reports*. Their work gave voice to a shared and growing frustration many had about maintaining the cars they owned, while it strengthened the unsavory and untrustworthy image of the auto mechanic in the public mind and kept criticism of the repair industry simmering into the third and strongest phase of the consumer rights movement (see fig. 26).

Addressing Congress in the spring of 1962, John F. Kennedy announced to a rapidly suburbanizing nation his "Consumer Bill of Rights": the right to safety, the right to be informed, the right to choose, and the right to be heard.[19] Presidential endorsement combined with extensive grassroots support gave consumer activists and elected officials considerable new social and political power in this third stage of the movement, which they aimed at numerous products and services. By 1968 a national survey by the Opinion Research Corporation found that "seven Americans in ten [thought] that present Federal legislation [was] inadequate to protect their health and safety. The majority also believe[d] that more Federal laws [were] needed to give shoppers full value for their money."[20] Legislators soon responded to consumers' automotive complaints, though their attention did not initially alight on the repair shop. Rather, consumer advocates and legislators first investigated automobile safety and automobile insurance.[21] Yet each of the House and Senate hearings on these issues touched on related problems in the repair industry. Hence, in 1968 Senator Philip Hart of Michigan convened highly publicized hearings before his Subcommittee on Antitrust and Monopoly of the Senate Judiciary Committee. The Hart Hearings, as they came to be known in the press, met sixteen times from December 1968 through

March 1970, generating over four thousand pages of published testimony and supporting documentation as well as untold numbers of media headlines and commentary. Hart was a low-key, deliberative, and liberal Democrat whom his colleagues referred to as "the conscience of the Senate." Intensely interested in the environment and consumer issues, he made his hearings the conscience of the auto industry by laying bare the problems that plagued consumers in the day-to-day use of their automobiles.[22]

Hart's subcommittee heard testimony from all sides of the auto repair problem, and the hearings provided a public forum for the frustrations that motorists, mechanics, dealers, and automakers experienced at the end of the 1960s. In the year following the first public meeting, approximately six thousand "let-me-tell-you-what-happened-to-me" letters found their way to Senator Hart's office.[23] Together with individual motorists' testimony at the hearings, they echoed the distrust captured in Riis and Patric's 1941 *Reader's Digest* "Gyp" articles, except that now the complaints were directed to the government in the expectation that something should and would be done. As Lawrence Roush of Wilson, North Carolina, explained to conservative senator Sam Ervin Jr., who also sat on the subcommittee: "[The automobile] is just as much a necessity for the American family as is the telephone, electricity, gas, and water. Why then does the Federal Government continue to pussyfoot around in this matter and permit the consumer to get gouged till it hurts? . . . The time for more stringent Federal regulation of the auto industry is long overdue."[24] The consumer rights movement gave state and federal governments new power to address motorists' frustration, and the Hart Hearings investigated how Congress might do so.

Witnesses at the hearings all agreed that a problem existed but differed in their assessment of its extent. Glenn Kreigel, owner of a diagnostic center in Colorado, set the critical tone of the hearings on the first day of public testimony. His firm only tested cars; it did not do any repairs. Customers could get a complete diagnosis of their cars' systems at his shop and take that information to a repair shop to have the work done. Many customers then came back to his shop for retesting to verify that the repairs had been completed satisfactorily. Kreigel testified that of the five thousand to seven thousand customer cars that his company retested, only about 1 percent had the repairs completed and correctly performed.[25]

T. A. Williams Sr., president of the National Automobile Dealers Association (NADA) took pointed exception to Kreigel's findings. He defended the industry, claiming that auto repair too often became "the victim of unjustified criticism" and that it had "frequently served as the 'whipping boy' for unqualified experts." He countered with statistics meant to soothe anxieties and to rebut the headline-grabbing testimony of Kreigel: NADA had done its own survey of ten thousand motorists and found 64 percent of them satisfied with their service work on the first visit to the dealer's shop—a far cry from Kreigel's damning assessment. Yet Senator Hart turned the ostensibly reassuring NADA figures around to ask Williams about the other 36 percent. "What happened?" Hart asked. Reluctantly, Williams admitted, "Chances are that a good many of the 36% did get poor service, We [automobile dealer service shops] are having our problems."[26] Whether inadequate auto repairs accounted for 36 or 99 percent of transactions in the shop did not change the general consensus at the hearings that things were bad in the industry.[27]

If quantifying the auto repair problem proved contentious, gauging the degree of threat it posed to society was also difficult. Although industry representatives carefully avoided saying that customers were wrong, some made it clear that they often found complaints trivial, cosmetic, or very difficult to trace based on customers' descriptions. The National Congress of Petroleum Retailers strongly opposed one of the actions under consideration, the creation of a consumer protection office at the federal level, because it "would only encourage much of the public to complain about minor incidences."[28] Some in technology's middle ground did not welcome an increased consumer voice.

Witnesses in favor of more thoroughgoing reform countered by playing the trump card of the consumer rights movement—safety—to underscore the urgency and gravity of the public threat. Senator Hart clarified, at least for rhetorical purposes, the priority of safety over the pocketbook issue of being "gypped" in the repair shop: "It is not just the additional cost to the automobile owner [that causes concern], but it is the additional exposure to injury and death."[29] Whereas Justice Cardozo had ruled in the MacPherson case over a half-century earlier that an automobile negligently made became a thing of danger, Hart and others argued that an automobile negligently repaired likewise became a hazard. Improper repairs to important steering, braking, and suspension systems could contribute to traffic accidents and even fatalities. Noting the number of mis-repairs and poorly maintained cars he had seen in his diagnostic center, Glenn Kreigel testified: "When I drive I start to reflect back on the day's work, and think, my God, we saw certain terrible, horrible things on a car. How many cars coming toward me might have that same defect and be over in my lane and I . . . would have no control over the situation whatsoever and become one of the count."[30] A few years later, before a similar investigative committee for the California State Legislature, Peter Carberry, executive vice president of the Automotive Service Council of

California, urged legislative action "to protect the lives on the highway. To get rid of the killer mechanic."[31]

The perceived threat posed by homicidal mechanics grabbed headlines and helped reformers gain political traction due to unprecedented public interest in automobile safety at the time. Consumer rights advocate Ralph Nader had been calling attention to auto safety since the late 1950s but made his biggest impact in 1965 with the publication of his book *Unsafe at Any Speed,* a scathing critique of U.S. automotive engineering. That same year Senator Abraham Ribicoff began a series of hearings investigating automotive safety. The following year Congress passed and President Lyndon Johnson signed the National Traffic and Motor Vehicle Safety Act, which mandated seat belts, impact-absorbing steering columns, padded dashes, and other safety features in all new cars. During earlier public debates about design safety, automakers sought to spread the blame around to poor drivers, to motorists who did not properly maintain their cars, and to mechanics who did shoddy work. Thus, for many considering the auto repair problem, poor repairs could be equally as dangerous as poor engineering and indeed could thwart the gains of safe design. In the eyes of auto repair reform advocates, conditions warranted federal intervention.

Intervention in the auto repair problem, however, required discovering its cause. Witness after witness before the Hart Hearings fingered various causes—most of them now familiar. Low wages, poor working conditions, and low status in the mechanic's occupation resulted in low recruiting and estimated shortages of "good mechanics" ranging from 40,000 to 150,000 nationwide. Furthermore, lacking certification or professional licensing, customers had no reliable, objective means to determine a mechanic's competence to do the work correctly. "License auto mechanics and you get rid of the butchers," claimed one correspondent to the subcommittee.[32] Witnesses berated flat rates as artificially inflating costs to the customer, while others defended them as the only way to keep labor costs under control. Still others maintained that flat rate–depressed wages were in fact the root of the problem. Others raised the accessibility issue, which had last gained attention in the 1920s. All were correct to some degree. Together they aptly described a diverse industry plagued with layers of social and technical problems.

While a range of witnesses paraded these well-known demons before the subcommittee, some identified an insidious new development from the mechanic's point of view—what John Kushnerick, editor of *Motor Age* magazine, described as "the mass of componentry" associated with the "antipollution system."[33] Harry Wright, a shop owner representing the Independent Garage Owners of America, warned Senator Hart and his colleagues that new troubles loomed on the horizon. "We know we are in for a lot of problems due to emission controls," he warned, "Emission controls breed problems." "The anti-smog situation has descended on the automobile industry," testified another. J. Howard Reed, of the Automotive Electrical Association, warned, "The Federal legislation on safety and emission control will only compound the problem."[34] Mechanics, shop owners, service managers, and their customers were just beginning to see the consequences of the environmental movement that, together with the consumer rights movement, would fundamentally alter the vehicles on which they worked. Looking at these changes from the vantage point of 1980, a Department of Transportation (DOT) study described "the development of a new type of automobile—the regulated automobile."[35] This complex machine would eventually challenge the entire sociotechnical system for creating and defining auto mechanics.

The Environmental Movement and the Regulated Automobile

While the environmental movement long predates the automobile, in the early twentieth century automobiles fostered the growth of modern environmentalism by forcing previously antagonistic wilderness and preservation groups to join forces in the face of threats that roads and noise posed to wilderness areas.[36] Even so, significant concerns about the environmental consequences of automobile tailpipe emissions did not surface until after World War II. Up to that point concerns about air quality centered on fixed-point sources: smokestacks, incinerators, and open fires. Regulators and citizens in industrial cities such as Pittsburgh used visible smoke density charts to monitor factory compliance with local smoke abatement ordinances.[37] Localized, episodic air pollution events such as the deadly smog incident that killed twenty residents of Donora, Pennsylvania, in the fall of 1948 kept much of the public and political attention focused on industrial sources of air pollution into the early postwar period.

Soon after the war, however, Los Angeles became the hothouse of automotive emissions control legislation and technology. As early as the 1930s, Los Angeles residents began to experience more than episodic smog events. The area's unique geographic basin, on-shore breezes, and flourishing oil refining industry convinced California legislators to authorize fixed-point abatement regulations before World War II—but to little effect. The state's postwar population grew at a rate far greater than the national average and coincided with a higher than average growth in vehicle registrations. Having controlled major fixed-point sources, common sense dictated to some that the automobile might be a significant source of the increasing smog. A 1949 football game at Berkeley finally pointed suspicion more clearly at the automobile.

On a clear fall day in 1949 the Cal Bears played the Washington State Huskies, and their fans drove from miles around to watch and cheer. They inched and idled their cars through jammed traffic to get to the game. As they did, a Los Angeles–like haze blanketed the vicinity of the game, and many experienced the stinging eyes and tight chest associated with LA smog. The extra automobiles seemed to be the only variables at work that day as they steadily pumped the exhaust fumes of thousands of small gasoline explosions every minute, hour by hour, into Berkeley's air.[38]

By then Arie Haagen-Smit, a CalTech biochemist whose prior research focused on isolating minute flavor compounds, had turned his expertise and laboratory equipment toward isolating the chemical compounds in Los Angeles–area smog. Since 1944 commercial growers of leafy crops such as spinach and lettuce had been experiencing heavy damage from the effects of air pollution, forcing some to move their farming operations or go out of business. John Middleton at the University of California agriculture research station in Riverside and Fritts Went at CalTech in Pasadena confirmed that the crop damage resulted from air pollution, but they did not isolate the pollution source. Haagen-Smit, Went's colleague and a member of the Los Angeles Chamber of Commerce's scientific committee, investigated the problem further at the chamber's request. From 1948 to 1949 Haagen-Smit filtered thousands of cubic feet of air surrounding CalTech's Pasadena campus, itself nestled near the foothills east of Los Angeles and often subject to atmospheric inversions that concentrated the smog. He determined that LA smog contained compounds that could only come from petroleum, though they did not remain in their original form. They had changed in the presence of sunlight into the irritating mix that had so long baffled observers. Haagen-Smit reported his preliminary theory on the photochemical formation of smog to the chamber and to the Air Pollution Control Board of Los Angeles County in 1950, confirming the suspicions of some that mobile sources—automobiles—were one of the major sources of the compounds that became smog.[39] Haagen-Smit's findings initiated a protracted political battle over automobile emissions which shaped the technology produced by American and international automobile makers for the remainder of the century.

By the end of the 1960s changes made under the hood of American cars indicated that the environmental movement had gained enough social and political power to set new boundary conditions within which Detroit engineers would have to operate.[40] This was new to Detroit. Up to that point the automobile manufacturing industry had enjoyed over a half-century of friendly, cooperative relations with governing bodies. Carmakers had benefited from generally favorable labor legislation, generous wartime contracts, depreciation and zoning laws that encouraged suburban sprawl and automobile dependency, dedicated taxes for building roads and infrastructure for their products, and generally cozy personal relationships with legislators in Washington and many statehouses. Now things were changing. They came under increasing public scrutiny and regulation, as did their products and those who worked on them after the sale.

The first under-the-hood manifestation of this battle appeared in 1961 model year cars sold in California. After delaying actions, Detroit finally succumbed to California legislative pressure in 1960 by installing positive crankcase ventilation (PCV) systems to capture and burn the potent mixture of fumes that slipped past the piston rings and collected in the crankcase. Automakers had long recognized that these vapors caused detrimental "oil dilution" in the crankcase, and most designed their engines simply to vent these unfiltered fumes directly to the atmosphere through a "down-draft tube" or similar device. These "blow-by" emissions accounted for up to a quarter of the total hydrocarbon emissions from automobiles, one of the key compounds Haagen-Smit later identified in photochemical smog. As early as the 1930s, automotive engineers knew that these fumes also presented a health hazard outside the crankcase, and in 1936 W. S. James of the Studebaker Corporation revealed a simple copper tube method for "recombusting" these potent fumes through the intake manifold. By the 1940s some Cadillac models employed a similar system for the purpose of "avoiding any possibility of unpleasant fumes reaching the interior of the car body."[41] Yet as late as 1959, the Automobile Manufacturers Association claimed that the PCV method for limiting air pollution "was brought to light recently by scientists of one of the automobile manufacturers."[42] Automakers then offered to install them "voluntarily" ahead of the California deadline and followed this move by adding them to new cars nationwide in 1963, a year ahead of federal mandates, in order to assuage growing concerns about automobile-fed air pollution. Thus began a cycle of distrust on the part of regulators. Detroit seemed to deny or delay its own technical abilities to design cleaner cars, only reluctantly responding to legislative imperatives. An antagonistic brinksmanship between regulators and automakers ensued for much of the remainder of the century.

Ultimately, the PCV system—a very simple technology with no moving parts beyond a check valve—added little to the maintenance demands of new cars in the early 1960s. Model year 1966 California cars, however, incorporated under-hood modifications that marked the beginning of the "componentry" that had bedeviled mechanics by the time of the Hart Hearings. In 1964 California's Motor Vehicle Pollution Control Board (MVPCB) certified four aftermarket exhaust treatment devices (three catalyst devices and one direct-flame afterburner) as meeting the emission reduction and durability targets established for new California cars. This triggered a prior legislative mandate that all cars sold in the state in the following model year, 1966, must meet those emission requirements. Again, Detroit delayed, remaining steadfast in its public protestations that such reductions could only be accomplished by the 1967 model year at the earliest. Yet in 1964, facing the threat of having to purchase, install, and warrant the aftermarket exhaust devices approved by the MVPCB and manufactured by other companies, each of the major Detroit automakers conceded that their own in-house efforts would, surprisingly, be ready for the 1966 model year. None used the devices developed by independent companies for the California market.

Chrysler chose to meet the California standards through its "Cleaner Air Package," which amounted to carefully tuning the engine to run cleaner. A heated air intake, retarded spark, and lean carburetor settings decreased carbon monoxide and hydrocarbon emissions to California standards while increasing oxides of nitrogen, which were not yet regulated. Chrysler's system also showed itself very sensitive to adjustments and maintenance; it could fall out of compliance relatively easily, and it was difficult to verify in-use compliance without testing equipment. General Motors, Ford, and American Motors initially met California standards with air injection systems ("Air Injection Reactor," or AIR, in GM cars; "Thermactor" by Ford; and "Air Guard" by American Motors) which pumped additional oxygen into the exhaust stream to facilitate continued oxidation of unburned hydrocarbons (see fig. 27). These systems also incorporated tighter air-fuel mixtures and retarded idle ignition but were more robust than Chrysler's and could be visually inspected for compliance. Air injection systems also cost more to manufacture. Thus, when the federal government adopted the 1966 California emission standards for 1968 cars nationwide, GM and Ford shifted to the cheaper, less stable Chrysler method on many of their cars.[43]

With the advent of the regulated automobile, motorists and mechanics began to notice new quality, "drivability," and maintenance problems. During California's brief failed attempt to mandate the retrofitting of all used cars with PCV devices in 1965, one historical study found that "automobile clubs, mechanics and used car dealers [claimed] that the devices were nuisances." Some vocal California mechanics feared "that the presence of devices would make ordinary repairs and maintenance more difficult, resulting in unhappy customers, the need to do jobs over, and so forth."[44] At public hearings about the used car law, the

MVPCB countered that the "only thing wrong with the devices . . . was old mechanics' tales, combined with bad mechanics' installations." After checking into complaints, the MVPCB found that "the devices had been installed incorrectly . . . upside down and backward."[45] Technical changes initiated by the environmental movement did not get off on the right foot with mechanics or motorists. The regulated automobile got a cool reception in technology's middle ground.

Further troubles ensued with the engine control methods employed in 1966 to meet California's tailpipe standards. In January 1967 California's MVPCB reported that 40 percent of 485 randomly selected 1966 cars it tested exceeded the 1966 emissions limits.[46] Even after less than two thousand miles of driving, 37 percent failed. Of those driven over two thousand miles, 63 percent failed, and at twenty thousand miles a whopping 85 percent failed.[47] In March of that year California smog officials warned automakers that if their emission controls were not "substantially improved, the cars would be banned from sale in the state."[48]

In reality the blame for poor emissions performance extended to all the major actors in technology's middle ground. The high failure rate indicated that automakers were either not putting their best efforts into manufacturing quality emission control devices or they were giving little weight to the long-term durability of those devices. Regulators and industry critics saw automakers as still dragging their feet and putting more effort and money into superficial design changes than they were into technological innovation or product quality. GM and Ford's shift to the Chrysler tuning approach in 1968 may have been a rational, competitive business decision, but it proved to many observers that their commitment to the nation's air quality was superficial and cheap.

Motorists shouldered some of the blame for the ineffectiveness of early emission controls as well. The American Automobile Association (AAA) of Missouri pointed to a number of things automakers could do to make cars better but also noted that most motorists "treat their cars as appliances, and give them about as much attention."[49] Automakers had argued from the beginning of the debate over tailpipe emissions that the problems lay not in the design of cars, but in the excessive pollution caused by poor owner maintenance. Thus, even when they designed cars to run cleaner, motorist's behavior could spoil any gains. Regulators agreed in part. While they did not absolve Detroit of its responsibility to produce high-quality, durable emissions systems, the propensity of some motorists toward procrastination and inaction concerning maintenance and repair would have to be addressed in the legislation and implementation of future emission controls.

In addition to neglectful motorists, tinkerers and performance enthusiasts presented another problem. As Detroit automakers emphasized tuning modifications in the late 1960s to achieve the desired emission test results, mechanically inclined motorists quickly learned to readjust those settings to increase the performance and drivability of their cars.[50] Furthermore, a 1970 federal study found indications that "a sizeable percentage of motorists do deliberately remove or inactivate their emission control systems."[51] Simply removing the belt driving the pump of a GM-style air injection system would render it inoperable. Two years later the owner of a 1972 Chevrolet Vega wrote to Smokey Yunick for help with constant stalling. Yunick replied that the problem stemmed from "the tight emission standards, which force the carb design to stay way over on the lean side."[52] Smokey's reply implied that the owner might want to think about enriching the idle fuel mixture. If the Vega owner did so, he would end the stalling problem but would also take the car out of emissions compliance. Automakers tried to inhibit such tinkering by incorporating physical limits to the idle mixture adjustment of their carburetors. Since the earliest days of motoring, the carburetor had been a contested under-hood site where designers attempted to prevent motorists and mechanics from making hasty adjustments that would affect the operation of the engine.[53] The carburetor struggle took on new urgency after 1966 because maintaining a precise air-fuel mixture going into the engine significantly reduced the pollutants that came out of the tailpipe. Indeed, following passage of the 1970 Clean Air Act Amendments, the Environmental Protection Agency's (EPA) Office of Enforcement and General Counsel formulated and distributed its Mobile Source Enforcement Memorandum No. 1a, which redefined under-hood tinkering that took cars out of emissions compliance as "tampering," a federal offense.[54]

If automakers and motorists could not be trusted with emissions control devices, neither could auto mechanics. William Megonnell of the National Air Pollution Control Administration (NAPCA) expressed the regulatory view when he pointed to studies showing that California mechanics actually made emissions worse after adjustments, indicating a "pressing need for many more, much better trained repairmen."[55] In response, Megonnell's agency began requiring 1970 automobiles to carry "conspicuous and permanent" under-hood decals providing the manufacturers with basic tune-up specifications as approved for emissions certification. The agency intended these decals "as an aid to mechanics in the proper adjustment of engines to reduce air pollution emissions."[56] Nevertheless, according to Miles Brubacher, chief engineer for the California MVPCB, "the ordinary run-of-the-mill mechanic just doesn't know what he is doing in relation to emission control. There has been a lot of training going on over the years since

1966, but it just hasn't done the job."[57] Frank Daley, director of Service Research for GM, endorsed Brubacher's assessment from the automakers' perspective. Daley candidly admitted to an audience of industry insiders in 1970 that, given the way they were designing cars to meet emission standards, "we all know today that the cars do not run exactly as smoothly as they did before. The idle is a little bit unstable and so forth as you set them according to the specifications that are in the manual." The way Daley saw it, "The old time mechanic wants that car to run right . . . so the mechanic has a tendency to smooth the engine out somewhat . . . a richer mixture [or] whatever it might take. And this is part of the problem today." Paul McKee, Emissions Programs manager at Ford, concurred and suggested showing mechanics that their customers will not notice the rough idle of an engine tuned to emissions specs because "engine mounts [made of thick rubber] take up a lot of the rocking."[58] Daley and McKee seemed to think that mechanics and motorists should be educated to accept rough engine idle as the new standard.

Dullness or duplicity, however, did not always explain mechanics' behavior. Not all mechanics acted as ignorantly or carelessly while working on emissions-certified cars as regulators and automakers implied. Like Smokey Yunick, they knew that the lean carburetor settings and retarded ignition timing decreased the performance and drivability of their customers' cars. Making customers happy meant improving a car's performance in ways detectible from the driver's seat, even if the result defeated the emissions goal of the manufacturer or government anti-pollution administrators. Attentiveness to their customers motivated them to "de-tune" emissions certified cars. Nonetheless, if emission controls were to remain effective, regulators would have to deal with mechanics' deviant behavior as well as that of motorists and automakers.

The Diagnostic Fix

By the time the Senate Subcommittee on Antitrust and Monopoly concluded its public hearings on the automobile repair industry in March 1970, the mutual distrust on all sides loomed as significantly as the assorted causes of the industry's troubles. The members had heard from just about every interest group with a stake in the outcome, but conventional legislative remedies seemed elusive. Hart conceded that training more and better mechanics would help yet made no specific recommendations beyond supporting existing public education, industry training, and federal Manpower programs. He took no action on the issue of flat rates, and he backed away from earlier remarks favoring the licensing or certification of mechanics. Instead, Hart placed his faith in technology—in the promise of diagnostic test centers—to fix the problems facing consumers in the repair shop. He sought to inject trust into technology's middle ground by separating diagnosis from repair and placing the former under the consumer's control.[59]

Hart's move away from mechanic licensing greatly disappointed many. He feared, as dealer and automaker representatives argued, that "licensing of all mechanics may cause more problems than it would solve—such as raising the overall repair bill." Constituents ranging from consumers to employers pleaded with him not to forsake certification. Even Donald A. Randall, assistant counsel to Hart's subcommittee, who was instrumental in orchestrating the hearings, called for a government program of licensing mechanics in a consumer exposé he published after the hearings. The initial threat of federal intervention, however, spurred the National Automobile Dealers Association to propose a voluntary mechanic certification program by the time the hearings drew to a close. NADA and the Automobile Manufacturers Association funded creation of the National Institute for Automotive Service Excellence (NIASE), which worked with the Educational Testing Service (ETS) to launch a nationwide program in 1972. By 1976 NIASE certified approximately eighty-two thousand mechanics in at least one of eight specialty areas.[60] A few states and some municipal governments still pursued mechanic certification or the licensing of repair shops, and many states passed specific auto repair disclosure laws and anti-fraud statutes aimed at curbing deceptive or crooked mechanics in the years following the Hart Hearings. But NADA's action eased pressure for more thoroughgoing reform of the mechanic's occupation.[61]

Rather than testing and regulating mechanics or shops, Hart proposed a "nationwide system of diagnostic testing centers run either by the States or by State-licensed private agencies [which] would not be associated with repair facilities." "Daydream with me a minute on the potential of it," he told a group of regulators and industry representatives. Such a system would coordinate with and improve state safety inspections by checking car owners' maintenance of their safety and emissions control systems—or their tampering and removal—thus preventing deaths, injuries, and further degradation of the environment. It could also keep tabs on automakers by amassing data on each of their products to track reliability, defects, and recalls. The gathered data could lead to periodic publication of the "weak points" of each make and model, encouraging manufacturers to improve their quality and empowering consumers and regulators with objective information about those who did not make improvements.[62] Hart even believed such a system could ease the auto repair industry's labor shortage because, "given a network of diagnostic centers which can pinpoint the problems scientifically we will be able to use lesser-skilled persons to do some of the repair work." Finally, and most important to Hart, customers could have their vehicles' problems diagnosed "before the owner turns himself into a garage or over to a mechanic," thus saving millions of dollars in unnecessary repair work each year.[63] This single technological reform, Hart believed, could cure a host of problems in the auto repair industry.

Hart backed this faith in the diagnostic concept by sponsoring the only significant legislation to emerge from his subcommittee's hearings. The Motor Vehicle Information and Cost Savings Act, signed into law by Richard Nixon in 1972, provided the secretary of transportation $92.4 million over six years to establish five to ten "motor vehicle diagnostic inspection demonstration projects." Diagnostic test centers flourished in the years surrounding Hart's subcommittee hearings. Hart's bill, however, aimed at refining and redirecting their development toward a system he believed would be more effective in curing the ills of the auto repair industry.[64]

Diagnostic test centers grew out of the same faith in scientific objectivity—and the merchandising utility of that faith—which had motivated the introduction of large, roll-around diagnostic units such as Ford's Laboratory Test Set in the 1930s.[65] These early test sets had received a cool reception from many mechanics, and by the 1940s the primacy of their sales function grew increasingly apparent to customers. Roger Riis and John Patric described them as "Rube Goldberg machines" in their series of "Gyp" articles for the *Reader's Digest*. "Efficient and modern pieces of testing apparatus they may be, as their makers claim," the journalists conceded, but in their study "they were *always* used to back up a crooked explanation of wholly imaginary defects."[66] Nonetheless, faith in the ideal of objectivity and the belief that what was wrong with a car could be objectively known, independent of the mechanic's skill, personality, or character, remained intact and gained new adherents in the 1960s. The new apostles of objectivity, including Senator Hart, vested their faith in this latest manifestation of science in the shop.

The tools of this new objectivity differed from the test units of the 1930s and 1940s. Rather than single pieces of equipment, diagnostic centers, or "car clinics," encompassed entire shop bays or drive-through diagnostic lanes where "technicians" evaluated the condition of the entire car—including brakes, lights, shocks, suspension, alignment, engine, and transmission. They utilized improved diagnostic equipment as well as an array of other mostly conventional measuring and testing equipment to conduct anywhere from sixty to over one hundred individual checks on each car.[67] In their most complete form diagnostic test centers included a dynamometer to test performance at highway speeds, and some centers employed computers to compare test data quickly with manufacturers' specifications and tolerances.

Mobil Oil Corporation opened the first retail diagnostic test center in Cherry Hill, New Jersey, in November 1962, and soon other oil companies, mass merchandisers, dealerships, and garages followed suit. Many believed such "dispassionately accurate" centers would finally change the industry for the better, and a mini-boom in center building ensued from the late 1960s into the 1970s. John Kushnerick, *Motor Age* editor, told the Hart Hearings that diagnostic centers "are providing a very definite pattern . . . for what we will see in the seventies." In fact, his magazine's census of diagnostic centers revealed a nationwide growth pattern from 82 centers in 1967 to 227 in 1968, 366 in 1969, and 533 in 1970. Call Carl's in Washington, D.C., installed its "Computer Diagnostic Center" in 1967, the same year that Ford broke ground on its Autolite Car Service Clinic in Springfield, New Jersey. By the end of 1970 J. C. Penney alone operated eighty "Scientific Test Centers" (see fig. 28).[68]

Yet as a business proposition, these centers proved a costly merchandising investment. A model two-bay diagnostic center for a Ford dealership cost an estimated $50,000 for the building, excluding land, plus another $35,000 in equipment. In addition, because it took fifty-five to seventy-five minutes for each car to wend its way through the battery of visual inspections, measurements, and diagnostic tests, operators found it difficult to price services high enough to cover costs but low enough to lure customers. Call Carl's increased its test price from $11.95 in 1967 to $13.95 per car in 1971 but incurred costs of $15 to $17 per test. The AAA of Missouri ran a test center for club members, charging $15 for members and $20 for nonmembers. Nevertheless, in 1971 it reported losing money on the center for two years running. Most shops, therefore, installed diagnostic test bays or lanes, expecting that a good percentage of test customers would also become profitable repair shop customers.[69]

Pressure to secure repair sales, however, risked tainting customers' perception of their objectivity. Nevin J. Rice, who ran a center in Jackson, Mississippi, confessed to a *Wall Street Journal* reporter in 1968: "Frankly, we use the hard sell. . . . We want their business and we let them know it." One equipment maker suggested that centers "hire pretty girls to break the news of repair estimates to customers." Many offered coupons, discounts, or rebates on the test fee if customers had the repair work done at their shops. Furthermore, despite their much-touted scientific objectivity, consumer advocates and investigative journalists exposed wide variations in test results for the same car at different centers, proving to one consumer organization "that the centers are not necessarily the answer to the motorist's prayers."[70]

Even some within the industry believed diagnostic centers might not be up to expectations. Their influence "is a very debatable situation in our industry today," shop manager Harry Wright told Hart's subcommittee. "I have had a number of my customers' cars coming in with diagnostic centers' reports, and . . . I have actually taken my customer to the car and shown him on diagnostic equipment where the report he received from the diagnostic center was not correct." James Hall, owner of "Red" Ivey's Automotive Service in Atlanta, also doubted their value. He had heard all the hype and wanted to add a center to his shop, so he studied the concept, visited diagnostic centers, and spoke with equipment manufacturers. "Now," he told the subcommittee, "I have reached the conclusion from everything that I have seen that there is a vast area of ignorance in getting the state of the art down to the preciseness that would warrant the expenditure of money."[71] Nevertheless, their criticism betrayed a core faith in the equipment's objective potential. Wright used his own diagnostic equipment to convince his customer that the diagnostic center's report was wrong. Hall questioned the current state of the art but did not rule out the possibility of developing sufficiently precise equipment in the future.

Senator Hart and other devotees of diagnostic test centers believed that their promised objectivity could be salvaged from the pressures of the marketplace. The technology was not the problem, just its deployment. If better diagnostic equipment could be developed, and if diagnostic centers could be separated from the business of repair, then customers could trust their results. It was no surprise, then, that Hart opened his subcommittee hearings with Glenn Kreigel's testimony that ninety-nine in one hundred cars he retested did not have work performed completely or correctly. Kreigel's Auto Analysts center in Denver was one of just a few test-only diagnostic centers in the nation, and Hart looked to Kreigel's experience as ideally objective.[72] The public image of the mechanic and the highly publicized abuses in the auto repair industry worked against trusting scientific equipment in mechanics' hands. "Today's diagnostic centers," Donald Randall, counsel to the subcommittee, wrote in 1972, "still require the human element in interpreting data. And they are not idiot-proof."[73] Hart, Randall, and others believed that the remedy lay in idiot-proof equipment and independent testing centers. "I doubt very much," Hart told an audience of industry leaders and government officials in the spring of 1971, "that the users of automobiles in this country, the consumers, would question the value of a nationwide system of diagnostic centers and periodic motor vehicle inspection." "Now, it won't happen by noon tomorrow," he continued. "It may even be as tough as getting out of Vietnam. But both things are going to happen. . . . In both cases, the sooner the better."[74]

Hart's Motor Vehicle Information and Cost Saving Act therefore sought to hasten the development of such a diagnostic system. It required the Department of Transportation's demonstration projects be conducted by or under state supervision and that at least half of them must be test-only centers not affiliated in any way with auto repair businesses. Furthermore, the funded projects should incorporate any existing state safety and emission inspections, provide sufficiently detailed diagnosis of any failed components in order "to facilitate correction," provide for reinspection after repair, and gather data on the effectiveness and costs of repairs. Finally, the law required that the demonstration projects evaluate the cost and ease of use of available diagnostic equipment and suggest design modifications for equipment and automobiles to facilitate rapid diagnosis.[75]

At its core the vision of advanced independent diagnostic centers pursued a technological remedy to the long-standing tensions between motorists and mechanics by replicating the opaque skill of diagnosing a car's troubles. Diagnostic testing would balance the asymmetry of technical knowledge in the repair encounter. Without becoming more knowledgeable about their own cars, motorist-consumers would gain control over the diagnostic process. "Ideally," NHTSA reported back to Congress, "a diagnostic inspection facility would use standardized and highly automated inspection equipment and data handling techniques to pinpoint the vehicle component(s) which caused failure. . . .The motorist would then take his vehicle to a repair establishment and relay instructions to the mechanic concerning the necessary work."[76] Expressing a sentiment reminiscent of early motorists' desire to rein in and instruct their chauffeurs, advocates of diagnostic test centers sought to assert control over the auto repair industry by circumscribing the authority of mechanics. The diagnostic center print-out would help put mechanics in the social position of servants, carrying out the motorist's orders of what to repair and, following the dictates of the flat rate manuals, how to fix it and how much to charge.

Yet could machines actually replace mechanics' diagnostic skills, their body-based synthesis of sounds, smells, vibrations, pulls, and bounces? Could they do the work of Smokey Yunick's ears? Some thought so. In 1968 Smokey went head to head with a Mobile Oil Company diagnostic center in East Meadow, Long Island, New York, for *Popular Science* magazine. "Good mechanics are getting tougher to find," Devon Francis wrote. "Can electronic brains take their place?" Smokey listened, smelled, felt, looked at, and measured all of the items that the diagnostic center checked on the same 1965 Ford. In the end he and the diagnostic center report agreed on seventy items and disagreed on twenty-one. The center missed some important issues, such as the engine's low compression, but flagged some things Smokey did not. Overall it came out essentially a draw and slightly cheaper for Smokey's diagnosis if charged at the going rate for mechanics in the area. On that result a bystander quipped: "Yeah? How are you going to get Smokey Yunick for $7.50 an hour?"[77] If a machine could approximate the diagnostic ability of a good mechanic, it would be a bargain and a relief to anxious motorists, even if it cost a bit more.

Many actively pursued ways to replicate a good mechanic's visceral knowledge. At a 1971 technical conference called to gather information on diagnostic technology, all of the speakers embraced the diagnostic concept. Diagnostic technology, they believed, would reduce error, ease customer apprehensions, and decrease the demand for highly skilled mechanics. Konrad Murrer, general manager of Call Carl's, described his company's diagnostic center vision, as "eliminating the human element whenever possible and replacing it with instrumentation." Lynn Bradford expressed NHTSA's "hope that we can translate from the subjective judgment area into the quantitative and definitive dynamic test so that the public can feel it's not somebody's opinion but a machine that can be calibrated."[78]

Fred Pradko of the U.S. Army Tank-Automotive Command reported progress working with defense contractor Dynasciences Corporation to develop a portable diagnostic unit that would collect sensor data on more than twenty-one engine parameters. Using the unit, he explained, the army mechanic "will not have to be concerned about this data because the diagnostic unit will analyze the data and make the decision for him as to what is wrong with the vehicle. . . . It will reduce the skill level that he needs to diagnose our vehicles." Pradko also described an already operational diagnostic center for evaluating tank engines and transmissions that had been removed from their vehicles. "No human intervention is allowed or permitted," Pradko said of the Multipurpose Automatic Inspection Diagnostic System (MAIDS). In addition to the usual temperature, pressure, flow, and voltage sensors, the MAIDS unit included twelve "valve cover accelerometers" and one "engine block accelerometer" to detect "internal malfunctions"—a baker's dozen of vibration sensors to automate the mechanic's sounding rod, to replace Smokey's ears.[79] In fact, army plans for the mobile diagnostic set actually reversed the flow of expertise. The mechanic would wear a set of headphones in order to receive instructions from the unit while performing a test drive. In practice the headset would shield the mechanic's ears from sounds emanating from the vehicle, replacing mechanical noises with the automated instructions from the diagnostic unit.[80]

Douglas Toms, acting administrator of NHTSA, admitted to conference attendees that he was "becoming somewhat enamored" with the concept of diagnostics: "It is my personal view that we are going to begin to see very quickly . . . [how] we can move to a better state of the art so that all people need to do is pull into the garage or pull into the certified testing station and a man opens a door and puts in the plug and throws the switch and away the thing goes, and that computer goes clickety-clackety and the guy gets all the information he needs."[81] Indeed, Volkswagen, had begun installing a standard diagnostic plug in its 1970 model cars which would enable their dealer service shops to connect a specially designed service bay diagnostic unit quickly. In short, diagnostic test centers could reduce employers' and motorists' dependence upon skilled mechanics, making it safe "to use lesser-skilled persons," as Hart had envisioned.[82]

The United States eventually pulled out of Vietnam, but Hart's vision for diagnostic testing did not come to pass. Despite Hart's efforts and the millions of dollars that the Department of Transportation, the U.S. Army, and the various equipment manufacturers invested in developing equipment for diagnostic centers, such a widespread independent system never materialized. The independent diagnostic center concept encountered a number of political, economic, and technical problems in the 1970s which hampered the efforts of even its most ardent supporters.

The legislative core of the concept, Hart's Motor Vehicle Information and Cost Saving Act of 1972, suffered from administrative neglect. Hart's act had emerged from the consumer rights movement, but with the inauguration of President Richard Nixon in 1969, consumer issues secured ever less White House attention. Such activism faded in the mid-1970s, as economic recession, unemployment, and energy costs took center stage in public dialog. Upon signing the 1972 act, Nixon was rumored to have told aides and Department of Transportation officials who were present "that he did not want any more of these—expletive deleted—consumer bills before him."[83] For the remainder of the Nixon and Ford administrations the Department of Transportation disbursed money to the five pilot projects under Title III but did not pursue Hart's vision of gathering, analyzing, and building on the information reported by those projects. Hart retired from the Senate in 1975 and died from cancer the following year.

In 1978, anticipating renewed support for consumer issues in the Carter administration, Representative Bob Eckhardt, a like-minded Democrat and friend of the late senator, opened hearings before his House Subcommittee on Consumer Protection and Finance on the subject of reauthorizing the Motor Vehicle Information and Cost Savings Act of 1972.[84] Eckhardt had shepherded the House version of Hart's bill toward passage in 1972 and used the new hearings to voice frustration at the lack of progress by the Department of Transportation over the previous five years. Donald Randall, Hart's key staff counsel during the earlier hearings and during drafting of the act, testified now as counsel to the Automotive Service Councils, Inc., successor organization to the Independent Garage Owners of America. After criticizing the Department of Transportation, Randall voiced strong support for reviving the 1972 act, restoring its intended consumer information provisions, and pushing the DOT to move forward with the diagnostic center pilot projects. Representative Eckhardt called Carter's new DOT administrator, Joan Claybrook, to testify before the subcommittee. Claybrook, a former Nader associate, might have been expected to embrace the original vision of the act.

A tense question and answer exchange between Claybrook and Eckhardt, however, left no doubt of her coolness to the idea of establishing an extensive, and expensive, network of independent diagnostic centers. Claybrook noted that under Title III "there were funds spent doing a lot of work," but that work had been badly organized and did not result in useful information. Her agency did not now have the funds to go back and fix the years of misspent efforts. Title III was not her baby, and Eckhardt's aggressive questioning of her about the actions of an agency she had headed for only three weeks did not dispose her to embrace it. The next year she instead told Eckhardt's subcommittee that her agency would be "looking closely at current and future technologies for on board sensors and diagnostic systems that could make automobile inspection, maintenance, and repair more accurate, efficient, and economical."[85] Claybrook still saw a technological fix to the auto repair problem, but the time for independent diagnostic centers had passed—both because of political inertia during the Nixon-Ford years and because of revolutionary technological changes occurring under the hood.

Diagnostics Get on Board

Although Hart's plan for diagnostic centers languished in the 1970s, the legislative efforts of his good friend and Senate colleague, Edmund Muskie, set in motion the under-hood changes that allowed Claybrook to embrace on-board diagnostics rather than off-board diagnostic equipment and centers. During the late 1960s Senator Muskie gained a reputation as "Mr. Pollution" in the Senate following passage of his 1965 Motor Vehicle Pollution Control Act. That act marked the beginning of federal involvement in tailpipe emissions and inaugurated five successive years of increasing air pollution-related bills introduced in Congress by members of both political parties. Following the giant oil spill off the coast of Santa Barbara, California, in 1969 and the legendary fire on Cleveland's Cuyahoga River, environmental issues of all stripes gained new importance in Washington. A 1970 poll revealed that 69 percent of Americans believed that air pollution posed ether a somewhat or very serious threat. The House and Senate both saw bills as severe as Senator Gaylord Nelson's proposal to ban internal combustion engines after 1975, while on 22 April 1970, the first Earth Day, students at the University of Minnesota ceremonially buried an automobile engine, and on other campuses demonstrators burned whole cars.[86] The public mood favored strong legislative action.

Responding to heightened public concerns, Richard Nixon used his State of the Union address that year to announce his plan for tighter automobile exhaust emissions and other measures. Senator Muskie, now a potential Democratic candidate for the presidency in the next election, was unwilling to work with the Nixon bill under consideration in the House. Instead, in late August he brought forward out of committee his own legislation featuring the Nixon bill's tighter standards but requiring that they be implemented five years sooner. Muskie's bill would roll back emissions severely in order to ensure the public health and at the same time accommodate projected growth in automobile usage. Muskie's bill won passage, and Nixon, on the last day of 1970 and without inviting Muskie to attend, signed it into law as the Clean Air Act (CAA) Amendments of 1970, the most sweeping federal environmental legislation to date.

Muskie intentionally drafted the CAA to force technological development in pollution control. He knew, as Hart's subcommittee had learned, that Detroit automakers regularly had dragged their feet on emission control technology and that their lobbying power could easily delay or derail efforts if Congress left the details to agency regulators. He therefore drafted specific automobile emissions reductions directly into the act. Automakers had until model year 1975 to reduce carbon monoxide (CO) and hydrocarbons (HC) emissions 90 percent from 1970 levels and until 1976 to reduce oxides of nitrogen (NO_x) by 90 percent. These reductions went beyond the technical capabilities of automakers at the time, and meeting them pushed Detroit both to lobby for delays—gaining a series of three one-year delays—and to innovate and introduce key changes under the hoods of their cars.

American automakers' decision to adopt catalytic converters as their primary means to achieve Clean Air Act emissions levels carried significant consequences for technology's middle ground. While automakers' intense lobbying secured delays in federal CAA emissions deadlines, California regulators held to their own 1975 deadline for HC and CO reductions. This forced automakers to turn back to the catalyst industry, whose products they had rejected back in 1964, when California approved three catalyst devices and triggered the 1966 deadline for that state's first round of emissions reductions. Automakers' choice to tune their engines carefully or add air injection systems rather than purchase catalytic mufflers for their 1966–67 California cars had caused many catalyst makers to leave the motor vehicle emissions business, but some continued to develop catalyst technology for special commercial vehicle uses and could in the 1970s offer Detroit two-way catalytic converters that used platinum and palladium to oxidize HC and CO.

Catalytic converters did not in themselves require much maintenance attention, but they required other changes. They had no moving parts, required no electrical inputs, and needed no adjustments, but they could be easily overheated and ruined if too much unburned fuel entered the exhaust stream. Poor carburetor adjustments, engine misfires, fouled spark plugs, poor ignition timing, and a host of other maintenance problems could spoil the converter.[87] Thus, using these devices forced automakers and regulators to look more closely at ways to design leaner engine operation into their vehicles and to limit mechanics' and motorists' access to adjustments still further. Simple idle screw stops would be insufficient. The EPA had already redefined the act of tinkering with emissions-related equipment as tampering, a federal offense subject to fines up to ten thousand dollars. Yet sanctions alone would not work either. In 1977 Eric Stork of the EPA observed in an *Automotive Industries* interview: "It makes very little sense to force the industry to design a car to be clean and then let them stick a big screw on the carburetor which you can reach with your fingers and give it half a twist and then the car's as dirty as if it were never designed to be clean. That just makes no sense. So . . . we expect, beginning with the 1980 models, to get some of those easy adjustments, especially the idle mixture and choke adjustment, off the cars."[88] Subsequently, EPA's Office of Mobile Source Air Pollution Control defined acceptable "inaccessibility" for 1981 model cars: "Using simple tools (defined as those commonly found in an individual's home toolbox, including an electric drill)," EPA inspectors should not be able to gain access to and modify the design parameters of carburetors within thirty minutes.[89] Putting two-way catalytic converters in the exhaust stream reinforced and built on upstream changes in engine design to enforce downstream changes in behavior. Environmental concerns expressed in legislation and designed into carburetors began closing the hood on consumer tinkering and complicating mechanics' tasks.

Meeting the CAA reductions in NO_x emissions proved trickier for automakers and more consequential for mechanics. Gasoline engines produce high NO_x emissions under load when combustion temperatures are high and under lean air-fuel mixture settings—the ideal mixture for reducing CO and HC emissions. Initially, automakers used various Exhaust Gas Recirculation (EGR) systems to mix inert exhaust gases with the incoming air-fuel mixture in order to cool the combustion temperatures and thereby reduce the production of NO_x. The mechanical EGR valves, however, exposed to the hot and corrosive environment of the exhaust system, proved troublesome in operation. Between 1973 and 1985 automakers recalled more than ten million vehicles for EGR-related problems.[90] Dealership mechanics thus had to deal with each of the recalled vehicles, and independent shops no doubt saw many more that were not covered by specific recalls.

EGR systems alone still could not meet the tightening NO_x requirements, so automakers, having begun down the catalyst path for CO and HC, searched for a catalyst that would act on NO_x. They discovered that NO_x could be reduced in the presence of rhodium. But the catalytic reaction required a fuel-rich environment—just the opposite of the fuel-lean environment required to oxidize CO and HC with platinum and palladium. Thus, in order to make a three-way catalyst work in an automobile exhaust system, the air-fuel mixture going into the engine would need to fluctuate back and forth constantly in a very narrow band between slightly rich and slightly lean, providing just the right conditions for each stage of the catalytic converter to do its work in succession. In order to accomplish this air-fuel balancing act, automakers borrowed microprocessors from the growing electronics industry and put them in charge of carburetion. By 1979 Motorola, Intel, and Delco Electronics all provided microprocessors and controllers to the automobile industry. A network of sensors fed information to the microprocessor about the oxygen content in the exhaust, air temperature, engine temperature, throttle position, and more. The microprocessor then signaled the carburetor (eventually replaced by more precise fuel injection systems) up to thirty times per second to add more or less fuel to the intake, thereby ensuring the three-way catalyst the correct fuel environment to reduce CO, HC, and NO_x to regulated levels.[91]

As impressive as the microprocessor's work with the air-fuel mixture sounded, in relation to its potential it was just loafing when given only this emissions task. Automakers began controlling more engine and accessory operations through the microprocessor, or "engine control unit" (ECU), and began developing self-diagnostic routines to check the operation of the various sensors and systems.

Developments along this path toward computerized engine control to meet emissions requirements allowed NHTSA's Joan Claybrook to tell Representative Eckhardt's subcommittee in 1978 that her agency was looking at on-board diagnostic capabilities rather than off-board diagnostic testing.

At the same time, the EPA struggled with how to test vehicles in use to ensure that they remained in compliance. The CAA of 1970 directed the Department of Health, Education, and Welfare to establish National Ambient Air Quality Standards and required states not meeting these standards to implement plans to achieve compliance. These plans had to include mandatory emission testing of in-use vehicles to ensure that poor design, poor repairs, or a motorist's tampering did not hinder a car's emissions compliance. Yet testing emissions of vehicles in use proved difficult. Motorists could not be expected to leave their cars overnight so they could be run through the entire twenty-minute cold start–to–warm-up sequence used to certify new car models. Stationary short tests using tailpipe sensors at idle and fast idle could check CO and HC but not NO_x, which was produced under load. Testing centers with dynamometers could test for NO_x under load but carried a high price tag for extensive state use. Furthermore, EPA administrators such as Ken Mills, technical advisor in the Office of Mobile Source Pollution Control, remained uneasy with the correlation between short tests and the original new car certification tests. Meaningful correlation "just isn't in the cards," Mills believed in 1971.[92] By the late 1970s the presence of on-board sensors and ECUs presented EPA with another option.

As early as 1970, some began to speculate that it might not be necessary actually to analyze a vehicle's exhaust in order to verify its compliance. GM engineer Roy Knudsen expressed "a feeling in some parts of the automotive industry that you really don't have to measure the pollutant at all. You just have to design a car that will pass the pollutant test and then have the car adjusted properly, and it will reproduce that model car that you approved." Frank Daley, also of GM, agreed: "If you restore a car to its initial condition, . . . if it complied in the first place, then you're home free."[93] In other words, checking the designed operating parameters of a vehicle might be a way around relying on sensitive and expensive exhaust gas analyzing equipment to verify in-use emissions compliance. Such musing meant little, however, until the decision to use three-way catalytic converters brought sensors and microprocessors on board. When combined with the prospect of proliferating electronics and powerful ECUs in the late 1970s and early 1980s, the parameter-monitoring approach to emissions testing paved the way for On-Board Diagnostics, or OBD, in the late 1980s.

On-Board Diagnostics, first required by California regulators in 1988, initially piggybacked on the data-handling capacity and control modules already largely in place to accommodate three-way catalytic converters. In its fullest elaboration to date, known as OBD-II and phased in nationwide from 1996 to 1999, the system came to utilize more than four dozen sensors buried in just about every major component and subsystem of the automobile (see fig. 29). Other than two oxygen sensors, none of these sensors actually sniffed the tailpipe gases, and none directly measured the HC, CO, or NO_x content of the exhaust. Rather, OBD systems monitored the operation and interactions of the components to ensure that they remained operating within their designed—and EPA-certified—parameters. When the OBD system detected deterioration or malfunctions affecting emissions, it set a standardized error code in the computer memory and, depending on the type of malfunction, illuminated the CHECK ENGINE light on the dashboard alerting the motorist to seek a mechanic's help. The computer's memory then stored a code that told the mechanic what to fix or at least where to begin looking.

By monitoring multiple engine operation parameters, OBD ultimately manifested the broader social distrust of automakers, motorists, and mechanics on the legislatively defined issue of emissions. In accordance with EPA regulations, the computer memory of OBD stored a record of malfunction codes to provide EPA regulators with easily accessible "in-use" data on the performance of the emissions systems that automakers design and install in their products. These codes, when downloaded in large batches from cars on the road for five thousand, fifteen thousand, fifty thousand, and even one hundred thousand miles, give a more realistic and complete picture of the quality and compliance of Detroit's—or Tokyo's or Stuttgart's—emission control technology. The same codes can easily indicate motorist tampering as well. Indeed, the computerization of automobile systems has in many ways closed the hood on a range of user modifications, both because of the legal sanctions against emissions tampering and because of the complex ramifications that tweaking one component or setting might have on any number of other computerized sensors and data streams feeding information to the ECU. Many automakers now purposely obscure the technical details in their engine compartments. Upon opening the hood of Volkswagen's 1999 Passat, one finds a black plastic shield, a sort of demi-hood, covering the intricacies of vacuum lines, ignition wires, fuel lines, and the like (see fig. 30). The engine compartment presents an aesthetically pleasing but cognitively impenetrable black box of a power plant. While not itself the product of computerization, the Passat's demi-hood symbolized the designed-in disincentives toward old-fashioned tinkering found in computerized automobiles.

Nevertheless, OBD failed to address the original consumer agenda of providing motorists with understandable and comparative technical information about maintenance and repair and in some cases made diagnostic information more opaque and more difficult for either motorists or mechanics to obtain. Illumination of the Malfunction Indicator Lamp (MIL), or the Check Engine Lamp, symbolized the failure of On-Board Diagnostics to remedy the asymmetry of knowledge between customers and mechanics. The MIL lamp tells the motorist almost nothing useful. A far cry from the diagnostic center print-out that Senator Hart had envisioned, the CHECK ENGINE light provides no indication of the nature of the problem detected. The narrowly defined emissions agenda of the EPA left the potential diagnostic and consumer information capabilities of the non-emissions functions of the ECU largely unregulated, non-standardized, and hotly contested in the courts, pitting automakers and their dealerships against independent repairers and aftermarket suppliers over access to the coding and programming of the of the on-board computer system. As Chrysler representative Frank Krich put it in 1992, the Clean Air Act Amendments of 1990—which mandated OBD-II nationwide—required automakers to provide all necessary information to diagnose and repair emissions systems, but the "*packaging of diagnostic routines in a format unique to Chrysler's vehicles is proprietary. . . . Requiring us to supply manufacturer-specific enhanced diagnostic tools to non-franchised technicians is clearly beyond the purview of the Act and may jeopardize the economic viability of our authorized dealerships.*"[94] Diagnosis would not be freed from the pressures of the marketplace.

Thus, the adoption of OBD systems did little to ease the tensions in the shop between customers and mechanics or to ease the demand for highly skilled mechanics, as Senator Hart's vision for off-board diagnostics had intended. Automakers made certain parts of cars vastly more reliable and durable by "hardening" their emissions-related systems in accordance with EPA-imposed "useful life" warranties of fifty thousand to one hundred thousand miles. In fact, most of the parts and labor included in a mid-1960s tune-up needed little or no attention in mid-1990s automobiles. Ignition points, carburetors, and even some distributors had been designed out of mid-1990s automobiles. Nevertheless, the hybridization of electrical and mechanical technologies introduced new types of problems—electronic failures such as glitches, phantom codes, and bugs which could be intermittent and nearly impossible to find using conventional diagnostic methods. On top of that, mechanics faced increasingly divergent, application-specific electronics and computers, even within the same automaker's product line. Rapid development along this path pushed an incredible array of systems onto the streets and into repair shops. Turning cars into rolling computers made finding good general mechanics even more difficult at the same time that it made motorists still less comfortable evaluating a mechanic's diagnosis and recommendations.

Automakers and aftermarket equipment makers struggled throughout the 1980s and 1990s to equip and train dealership mechanics and independent shops with yet another generation of test equipment and increased training to help them diagnose and repair the new computerized automobiles hitting the market. General Motors employed Intel's 286 processor in its Computerized Automotive Maintenance System, introduced to dealers in 1986; a few years later Chrysler developed its Mopar Diagnostic System, and Ford pushed its $35,000 Service Bay Diagnostic System. Each automaker sought to reassure its dealership shop managers that by using this new equipment, "You can return your customers' vehicles with the Peace of Mind℠ that they have been repaired efficiently and correctly."[95] The Snap-On tool company promised independent shop owners that its Vantage Power Graphing meter with "powerful glitch capture ability," would save "time, money, and a lot of grief."[96] Yet the auto repair industry soon found itself facing a serious challenge. "The increased use of computers on cars and trucks," wrote the service editor for *Automotive News* in 1991, "has changed the rules in the problem-diagnosis game. No longer are mechanics able to solve problems solely by the seat of their pants."[97]

Smokey Yunick, America's iconic seat-of-the-pants mechanic, began to notice this shift in the mid-1980s. Idaho motorist Chris Hutchinson grew frustrated with his 1982 Chevrolet Citation because it would occasionally hesitate or cut out as if the ignition were turned off—sometimes during a long drive, sometimes when going uphill, but not always. The problem persisted even after several trips to the shop "for a fuel-pump-pressure test, a new computer, a rewiring job, and more, all to no avail." So, Hutchinson wrote to Smokey for help. "My answer," replied Smokey, "will demonstrate how complex and frustrating things have become for today's backyard mechanic. (I had to get help from an engineer to figure out your problem.) From here on the public must realize that, given the intricacies of today's cars, finding a repair shop with the proper equipment and know-how is the car owner's biggest challenge."[98] Yet intricacy or complexity alone did not account for the turning point Smokey noticed. Automobiles have always been intricate and complex relative to their time. Rather, Hutchinson's 1982 Chevrolet Citation used an Engine Control Unit, electronic spark timing, and electronic fuel injection, none of which lent itself to traditional sounding rod–style diagnostics. Such increasingly computerized automobiles brought with them new repair epistemologies, new ways of knowing the car and its subsystems. By the end of the twentieth century automobile mechanics needed to be equally adept at ferreting out bad sensor data and computer glitches as they were at listening for knocks and hums. They needed to be conversant with both analytical and visceral technologies to a degree only hinted at in Ford's Laboratory Test Set experience in the 1930s. These were the changes that "frustrated" the backyard mechanic and made finding mechanics with the "proper know-how" doubly important at the end of the twentieth century.

What began as an effort to develop off-board diagnostics and eliminate the human element in the repair shop turned out quite differently when the electronics went on board. Automotive electronic systems retained key monitoring and surveillance features born of distrust, yet they placed new premiums on highly skilled mechanics who could combine visceral, mechanical knowledge with more abstract electrical diagnostic knowledge. Regulated automobiles required mechanics who could understand both the pistons and the computers under the hood, exacerbating the need for skilled mechanics—precisely the opposite of what many of the agents contributing to the development of diagnostic technologies intended. This change in the nature of the automobile—born not simply of advances in automotive engineering but of multiple social, political, and technological forces and contingent opportunities—challenged the nearly century-old sociotechnical system of defining and creating mechanics.

Computerized automobiles, in fact, brought together two socially divergent technological cultures—one centered on automobile repair, the other on electronics—and in doing so began to force open the formerly closed sociotechnical ensemble that had grown up around the mechanic's occupation. Particularly after World War II, young workers trained to work with electrical circuits, electronics, and eventually computers increasingly found jobs as "technicians." Not quite "white-collar," these workers still worked with and often repaired machines, yet they were not "blue-collar" workers because they usually had relatively high levels of education, worked with symbolic data, and did not often get sweaty or greasy.[99] When automakers, legislators, and agency regulators pushed the two technologies into a single artifact, they created a situation in which auto mechanics would need to become conversant in two technological cultures.

"We've got a problem here," General Motors chairman Jack Smith told automobile dealers in the late 1990s. "Education programs are not turning out graduates for the jobs of the future."[100] He had it partially correct. An industry group known as the Coordinating Committee for Automotive Repair (CCAR) got closer when it attributed a shortage of sixty thousand qualified mechanics to three factors: increasingly complex cars; training that has not kept pace with that complexity; and the perception by students, parents, and career counselors that auto repair is an "undesirable career choice." Young men and women with the analytical abilities and training required to understand and repair computerized automobiles could easily choose other, higher-status careers. Said Sherman Titens, president of CCAR, "We are competing for the same worker that might go to IBM."[101] Computerization of automobiles represented possibly as significant a technological development in the transportation status quo as the horseless carriage had in the late nineteenth century. Deeply entrenched social structures for creating and valuing auto mechanics appeared at the end of the century not capable of meeting the needs of the next generation of personal transportation.

Servants or Savants?

Revaluing the Middle Ground

In the late 1990s neurologist and writer Frank Wilson noted the similar, highly developed manual dexterity of a surgeon and a world-class sleight-of-hand magician and then went on to observe parallels in their audiences' perception of their skills: "The patient in a doctor's office or in a hospital and the person in an audience watching a magic show . . . participate in a ritual shifting of power and responsibility to another. Conceding helplessness, the patient says to the doctor, 'I trust you. I know you can heal me.' The magician is placed on the same kind of pedestal, even if it is only theater. For just a little while he is clairvoyant, wise, and strong. He contains powerful knowledge and can work magic."[1] Yet what happens when we replace the doctor or the magician in this performance with an auto mechanic? (See fig. 31.) The latter, too, displays highly developed dexterity and skill, and the temporary shift of power in his favor remains. But the audience's unwillingness to trust the mechanic-practitioner converts submissive expectation of healing or pleasant entertainment into fretful anxiety about getting gypped. This book has attempted to understand this puzzle of technology's middle ground by examining, side by side, the technological and social developments that have shaped the auto mechanic's occupation over the last century.

The history of the auto mechanic's occupation has been, at root, a story of the creation and maintenance of sociotechnical hierarchies. Such hierarchies are not exclusive to auto repair or to technology's middle ground.[2] Yet studying the auto mechanic's occupation provides new vistas onto their complex formation, their institutionalization, and their consequences.

Looking at very early auto repair confirms that significant technological change can be socially disruptive. The introduction of the horseless carriage upended the established social arrangements of personal transportation which had grown up around horse-drawn vehicles by the late nineteenth century. The disruption came not just from the novelty of the technology or the scarcity of experience with it but also from the mechanical complexity of the machinery and the distinction between automotive knowledge and animal husbandry. The reign of chauffeur-mechanics proved brief as wealthy urban motorists and their allies drew on their access to deeper, more powerful social structures—courts, legislatures, training programs, surveillance—to regain control and reestablish social order in their favor.[3] Still, despite the decisive downward turn in the chauffeur's status, enthusiasm for the new technology kept many Americans clamoring for mechanical knowledge of automobiles. Widespread use of automobiles—whether by middle- and working-class Americans or by the U.S. Army—necessitated molding that enthusiasm into an occupation.

As the auto mechanic's occupation emerged out of multiple sources and niches in the economy, further technological development of automobiles offered fewer opportunities to challenge social norms than had their initial introduction. Some women and African Americans employed automobiles to challenge gender norms and racial stereotypes, but the work of maintaining and repairing automobiles instead grew ever more tightly entwined with prevalent social hierarchies and institutionalized into the status quo of American society. Gender, race, and class segregation in military training, public education, and employment mingled almost inextricably with the visceral nature of early automotive technology, setting auto repair off as one of a number of "culturally segmented epistemological domains."[4] This process gave particular social meanings to specific ways of knowing and interacting with automotive technology. It reflected and reinforced the mechanic's position between producers and consumers of automobiles as well as vis-à-vis other occupational and social groups. The resulting sociotechnical ensemble gained considerable power and momentum by mid-century, and the stigma that society attached to the auto mechanic's occupation led in large measure to the crisis many perceived in the industry by the late 1960s. That crisis in turn contributed to the conditions favoring the development of diagnostic equipment, On-Board Diagnostics, and computerization.

Studying the history of those who have repaired cars, as opposed to the history of the Ford Motor Company or the impact of the Model T on rural America, highlights developments that neither production- nor consumption-centered frameworks of analysis can explain well. Such frameworks certainly generate relevant questions and explain many actors' actions in the automotive service industry.[5] Ford, for example, designed its Laboratory Test Set of the 1930s specifically to increase service sales at its dealerships—a perfectly understandable action when viewed within conventional analytical frameworks. The failure of the Test Set cannot be explained, however, by Ford's misreading of service customers' desires nor by consumers' rejection of the "Rube Goldberg machines." Neither do sufficient explanations emerge from the retail, wholesale, or production domains of the automotive marketplace.

The failure of the rash of diagnostic test equipment introduced in the 1930s can be understood only by paying attention to technology's middle ground—that area between production and consumption where workers maintain and repair artifacts that they do not create or own—and by understanding the particular ways mechanics gathered information and diagnosed problems, the values society placed on their type of knowledge, and the social institutions that helped maintain the knowledge-identity relationship of the mechanic's occupation over time. Just as invention studies alone do not satisfactorily explain technological change, neither are user studies the end of the line of inquiry. Artifacts' meanings do not end with their creation or their use. As we have seen, their repair is also fraught with meaning.

Focusing on technology's middle ground illuminates the diverse sources of technological change in fresh ways. We can now recognize, for example, that the computerization of automobiles did not represent the simple unfolding of technological and engineering progress, which thereby forced changes in the repair shop and the home garage. Nor can we accept automotive computerization as simply "technological development oriented toward a positive consumption decision."[6] The idea of automated diagnosis long appealed to automakers, but the network of dealerships and independent repair shops shielded manufacturers from the direct costs and benefits of repair. Thus, they lacked the same motivation to automate service that they felt in production. Conversely, automobile dealers, independent mechanics, and equipment manufacturers exercised little influence over the design of the product that they needed to diagnose and repair. Social as well as organizational distance kept any such influence to a minimum. The U.S. military remained the only entity with enough direct labor costs and with some control over the design of its vehicles to move ahead on automation of diagnostics apart from external pressures. Civilian development and application of automotive diagnostic technologies resulted, instead, from multiple sociopolitical pressures—the confluence of environmental and consumer movements, the regulatory imposition of new boundary conditions for vehicle performance, and the individual actions of motorists, mechanics, politicians, state and federal regulators, military defense contractors, and others. All of this activity coincided with technological developments and opportunities in the automotive, equipment, and electronics industries from which actors chose particular tools, innovations, and variations.[7] Studying the maintenance and repair of technology can thus provide rich narratives that bridge the "macro" deterministic view that technological change drives social change and the "micro" social constructivist view that human choices determine which technologies get developed.[8]

Just as important, rich middle-ground narratives can help highlight that as technological change occurs, its significance depends on the values and structures of meaning woven around particular qualities of artifacts. The realization that computerization in the late twentieth century represented a major challenge to the status quo, whereas the introduction of automatic transmissions in mid-century did not, comes from understanding both the nature of mechanical and electronic technologies (the stuff mattered) and the social values attached to each over the previous century (human relationships also mattered). Because technology's middle ground is thick with social meanings, historical agents, overlapping institutions, and material artifacts, we can follow the complex sources of technological change and sort the consequential from the merely developmental.

Like the early horseless carriage, automotive computerization presented, and continues to present, opportunities to disrupt the status quo and to rethink the middle ground. Not that mechanics are soon going to be taking Cadillacs with On-Star navigation systems out for joyrides. Instead, young men and women protest in quiet abstention by not becoming mechanics, by not clamoring for more auto shop classes in their high schools, and by not protesting when auto shop is removed to make room for another desk-filled classroom. Efforts since the 1960s to engineer improved reliability and diagnostic capabilities into automobiles have been good and valuable to consumers in many ways. Ignoring and undervaluing the person who does the repair, however, perpetuates the core problem of the service exchange.

Automakers and others are now putting considerable effort into upgrading the auto mechanic's image. Together with parts suppliers, tool suppliers, dealer associations, and educators, they formed various coalitions in the 1990s to combat the problem of creating high-tech mechanics capable of diagnosing and repairing problems with today's "rolling computers." General Motors and Chrysler joined forces in "a national campaign designed to get local dealers and high schools together." Known as the Automotive Youth Educational Systems (AYES) program, it promised cars, parts, and manuals to eligible schools, advanced training for teachers, and multimedia "career awareness packages" for students. According to GM's Stan Moore, "We walk down the halls of the [participating high] school to make sure the auto lab looks like the computer lab because image is a problem. Most people today still think 'automotive technician' means grease monkey, and the requirements today don't support that image."[9]

Nevertheless, the occupation remains in the ambiguous social space between production and consumption. Scholars who have studied the impact of computerization in other settings have observed that lines of work newly "infused with technical content by microelectronics generally have previously existing identities and statuses. Because those who do not do the work are unlikely to appreciate how the work has changed, perceptions of the technicians are likely to be constrained by existing cultural frameworks."[10] This raises some important questions. How culturally constrained is the auto mechanic's identity and status? Is the middle ground itself, the professional act of repairing, impervious to social acclaim, to new social valuing, to changed sociotechnical hierarchies? Will the repair shop of the future look like the grungy robot repair shop depicted in *Star Wars?*

The too-easy distinctions that our culture has created between manual and mental work, between "skill" and "intelligence," are inappropriate and limiting in the repair shop as well as in many other occupations.[11] Manual dexterity, the skilled use of our first digital tool, has been central to the evolution of the human brain. As a species, we would not have our intelligence without our skill. Furthermore, as individuals, our cognitive development depends on the early exercise of our hand-brain connection—that is, on using our visceral knowledge.[12] As changes in automotive technology force the rejoining of previously segmented epistemological domains within the repair shop, mechanics need not be either visceral or analytical; they need to be both.

Other case studies have shown that computerized automation in production settings could be ill managed and employed in ways that reinforce hierarchical domination and control "at the expense of developing knowledge in the operating workforce." Or the same developments could help collapse dysfunctional distinctions between "white-collar" and "blue-collar" work, improving worker satisfaction and operational efficiency.[13] Concerned actors in the automobile service industry need to think about how they might replicate the latter effect in the repair shop if they hope to attract highly qualified young applicants. In an important study of blue-collar workers in Boston in the early 1970s, sociologists Richard Sennett and Jonathan Cobb reported: "One of the saddest encounters we had was with a philosophically-minded auto mechanic. A part of him recognizes that he is 'deep,' as his friend puts it; yet he cannot really accept the fact of his intelligence. . . . For, if he is intelligent, why is he a 'grease monkey'? . . . [I]t is less painful [for him] to think he 'isn't much, just part of the woodwork,' than to respect his own mind."[14] The sociotechnical hierarchies that perpetuate such "hidden injuries of class" must change if we are to keep our computerized cars in operation.

Yet in boosting the mechanic's image, neither can policy makers neglect the need for hands-on experience. As proprietary electronics and legal "tampering" sanctions discourage under-hood tinkering and as academic subject assessments push shop classes, art studios, music rooms and other sensory motor subjects from public school curricula, will our social reserves of visceral skill sets wither and grow scarce? Or will we expose children of all classes and genders to the qualities of the material world, old and new, clean and dirty, mechanical and electrical?[15]

Mechanics, for their part, have begun to respond to the pressures and opportunities that computerization affords. Increasing numbers each year invest personal time to study for, take, and pass the certification tests offered by the National Institute for Automotive Service Excellence, now known as ASE certification. The motivations and rewards that they feel for doing so, both personal and professional, warrant further investigation. Lacking any other nationally recognized professional marker of their competence, ASE certification—often required for those working on emissions systems—may gradually help accomplish what earlier efforts of organized labor failed to achieve.

Mechanics, and those who train and hire them, could further boost the occupation's esteem by adopting the health care profession's current focus on improving communication with clients. R. Paul Robb, of the California Dealers' Association, told members of the California Legislature in 1974 that a factor contributing to the service problem was "poor communication to repair people of the true nature of the problem." Typical of industry insiders, Robb placed the blame on the motorist: "Customers tell us that a noise sounds like something—it comes from somewhere back there—or that it just does it sometimes, but they don't know when, they don't know why, or that famous line that all personnel know, 'My husband says—.' "[16] Mechanics know sounds. They work with them day in and day out. Understanding sounds is part of their magical power. Among themselves mechanics often ridicule motorists' feeble attempts to communicate their cars' problems. Tom and Ray Magliozzi, hosts of National Public Radio's weekly *Car Talk* show, make great comic use of this situation. They attempt to get listeners to mimic, on air, the sound their car is making. The resulting vocalizations are greeted with laughs and good cheer before they go on to try to diagnose the caller's problem. And like many mechanics discover when they return the customer's car, when Tom and Ray call the listener back a few months later to check on their diagnosis, they find they were wrong and joke that the motorist had made the sound wrong, so their diagnosis was wrong.

Health care workers likewise rely on patients' descriptions of their own symptoms. And the properties that patients describe, such as degree and location of pain, are no easier to verbalize than unusual engine noises. Health care workers, however, cannot so easily excuse a misdiagnosis by blaming the patient's inability to describe their symptoms accurately. Doctors cannot expect their patients to study up on physiology and gross anatomy in order to better convey their symptoms. Neither can representatives of the automobile service industry continue to grouse about motorists' ignorance of their machines. They need to look at how to make mechanics better at communicating with customers, better at listening, better at asking good questions, and better at respecting customers' ignorance about the technical aspects of their car while not wielding their own expertise as a weapon of class justice or racial or gender privilege.

Perhaps mechanics are beginning to think of themselves in a new light. Like surgeons began to do in the late eighteenth century, individual mechanics are increasingly wearing gloves while they work (see fig. 32). This may seem innocuous to the outsider and even some insiders, but gloving the mechanic's hand is both practical and symbolically important at this stage of the occupation's history. Over the years mechanics have wantonly doused and immersed their hands in a whole range of petroleum-based solvents and liquids. New materials now make it possible to manufacture gloves that resist the oils and fluids of their day-to-day work. A better understanding of the occupational health risks posed by many of these materials has led the Occupational Safety and Health Administration regulators and insurance carriers to demand that mechanics wear gloves, and mechanics themselves are becoming more receptive to using them. Cushioned knuckles and heat-shielding panels in palms and fingers turn gloved hands into more versatile digital instruments than ever before. All the while, gloves symbolically allow mechanics to protect their hands from excessive dirt and abuse. And at the end of the day wearers can shed the telltale grime and scars of blue-collar work. The high-tech glove, then, represents the tentative grasp mechanics may now have on middle-class respectability.

Significant technological change has thus created pressures and opportunities to renegotiate the social markers of the auto mechanic's occupation in the early twenty-first century. It is unclear just how well the range of actors and institutions involved in creating mechanics can integrate hand and mind in a sociotechnical system that values repair sufficiently to attract a wide range of applicants. What knowledge, training, education, status, working conditions, management relations, and pay structures will be appropriate to attract qualified candidates in sufficient numbers to keep our twenty-first-century personal transportation system operating smoothly? Will a growing social awareness of the finiteness of our planet's resources encourage us to place more value on extending the useful life of our artifacts? Will this lead us to value the workers who maintain and repair our technology more highly? All that is certain is that workers in technology's middle ground will be called upon to repair our vehicles—whether they are powered by gasoline, biodiesel, hybrid power plants, or hydrogen fuel cells. They will all need to be fixed because cars break down. They always will.

Introduction • Technology's Middle Ground

1. "A Stitch in Time, Etc., as Applied to Automobiling," *Horseless Age* 8, 8 May 1901, 125–26.

2. Robert Bruce, "The Place of the Automobile," *Outing Magazine* 36, October 1900, 65. *Outing Magazine* was not an auto industry trade publication but a sport and leisure magazine aimed at upper-middle-class readers.

3. This survey may even underestimate the number of motorists who do not work on their own cars because the readers of *Popular Mechanics* are more likely than most to do their own work. "Owners Reports," *Popular Mechanics,* 15 February 1994, 169–211.

4. Thomas Acton and Gary Mundy, eds., *Romani Culture and Gypsy Identity* (Hatfield: University of Hertfordshire Press, 1997); David Mayall, *Gypsy-Travellers in Nineteenth-Century Society* (Cambridge: Cambridge University Press, 1988); Pierre Claude Reynard, "Unreliable Mills: Maintenance Practices in Early Modern Papermaking," *Technology and Culture* 40 (April 1999): 237–62; Donald Sharp and Michael Graham, eds., *Village Handpump Technology: Research and Evaluation in Asia* (Ottawa: International Development Research Centre, 1982); Rama Lakshmi, "Mechanic 'Sir' Now a 'Ma'am': Low Caste Indian Women Repair Village Water Pumps," *Washington Post,* 5 December 2004.

5. Consumers and users who mend or tinker with their own artifacts may, in the eyes of some scholars, be participating in acts of production, blurring the line between production and consumption. But it is not until they perform this work for pay on technology owned by others that their activities bring them into technology's middle ground. Ruth Schwartz Cowan, *More Work for Mother: The Ironies of Household Technology from the Open Hearth to the Microwave* (New York: Basic Books, 1983); Nina Lerman, "From 'Useful Knowledge' to 'Habits of Industry': Gender, Race, and Class in Nineteenth-Century Technical Education" (Ph.D. diss., University of Pennsylvania, 1993); Aldren A. Watson, *The Village Blacksmith* (New York: Thomas Y. Crowell Co., 1968); Jeannette Lasansky, *To Draw, Upset, and Weld: The Work of the Pennsylvania Rural Blacksmith, 1742–1935* (Lewisburg, Pa.: Oral Traditions Project, Union County Historical Society, 1980); Kenneth Dunshee, *The Village Blacksmith: A Story of His Metal and Mettle* (Watkins Glen, N.Y.: Century House, 1957).

6. Susan Strasser, *Waste and Want: A Social History of Trash* (New York: Metropolitan Books, 1999); Mike Rose, *The Mind at Work: Valuing the Intelligence of the American Worker* (New York: Viking, 2004); Stephen R. Barley and Julian E. Orr, eds., *Between Craft and Science: Technical Work in U.S. Settings* (Ithaca, N.Y.: Cornell University Press, 1997); Julian E. Orr, *Talking about Machines: An Ethnography of a Modern Job* (Ithaca, N.Y.: Cornell University Press, 1996).

7. On the limited scholarly studies of the auto repair industry, see the Essay on Sources.

8. The mechanic's occupation is not stigmatized to the degree Conrad Saunders describes for kitchen porters and sweepers, but as we will see, the occupation's early and close association with servile status and livery is significant. Saunders, *Social Stigma of Occupations: The Lower Grade Worker in Service Organizations* (Westmead, Eng.: Gower Publishing, 1981).

9. Microsoft Works version 4.5 running on this author's computer includes a "task wizard" feature offering a form letter entitled "Your mechanic tried to rip me off"—indicating both widespread distrust and the distinction between "my mechanic" and "your mechanic."

10. Lesley Hazleton, *Confessions of a Fast Woman* (Reading, Mass.: Addison-Wesley, 1992), 120–24.

11. John Seely Brown and Paul Duguid credit psychologist Jerome Bruner for this distinction between "learning about" and "learning to be." Brown and Duguid, *The Social Life of Information* (Boston: Harvard Business School Press, 2000).

12. See John A. Jakle and Keith A. Sculle, *The Gas Station in America* (Baltimore: Johns Hopkins University Press, 1994); John Margolies, *Pump and Circumstance: Glory Days of the Gas Station* (Boston: Little, Brown, 1993).

13. This passage draws on the title of Walter Vincenti's study, *What Engineers Know and How They Know It: Analytical Studies from Aeronautical History* (Baltimore: Johns Hopkins University Press, 1990).

14. Michael Polanyi, *The Tacit Dimension* (London: Routledge and Kegan Paul, 1967); Douglas Harper, *Working Knowledge: Skill and Community in a Small Shop* (Chicago: University of Chicago Press, 1987).

15. While the term *skill* shows up occasionally in this study, I have tried throughout to refer to the knowledge and abilities that various workers brought to the repair of automobiles as technological knowledge, a phrase that allows for broad interpretation. It avoids the unnecessarily strong male gender connotations of the term *skill* and its narrow association with dexterity and craftsmanship. Using technological knowledge also avoids the implicit hierarchical connotations of *technical knowledge* and its association with "higher-level" engineering knowledge. Automobile mechanics' knowledge included dexterous skill as well as abstract technical knowledge derived from the principles of internal combustion. It included visceral sense-based knowledge as well as knowledge of human relationships, which at times could be of equal importance in their work. When thinking about auto repair, I have found it more useful to set aside the labor historians' well-trod debate over skill, deskilling, and shop floor control and think afresh about what mechanics know and how they know it. This discussion of skill versus technological knowledge is cogently and clearly addressed in Nina E. Lerman, "From Useful Knowledge to Habits of Industry: Gender, Race,

and Class in Nineteenth-Century Technical Education" (Ph.D. diss., University of Pennsylvania, 1993). Two classic arguments about technology's role in de-skilling workers in the production setting are Harry Braverman, *Labor and Monopoly Capital: The Degradation of Work in the Twentieth Century* (New York: Monthly Review Press, 1974); and David Noble, *Forces of Production: A Social History of Automation* (New York: Oxford University Press, 1984). Amy Sue Bix examines industrial and labor responses to the related debate about technological unemployment in *Inventing Ourselves Out of Work? America's Debate over Technological Unemployment, 1929–1981* (Baltimore: Johns Hopkins University Press, 2000).

16. Andrew Abbott, *The System of Professions: An Essay on the Division of Expert Labor* (Chicago: University of Chicago Press, 1988), reveals the centrality of diagnosis, treatment, inference, and abstraction in forming professional identities. Bonalyn Nelson describes the mechanic as an amalgamation of expert and nonexpert service worker, leading to ambiguous status. Bonalyn J. Nelsen, "Technological Change and the Rise of the Service Economy: Observations from Automotive Repair" (Ph.D. diss., Cornell University, 1998).

17. See Harry W. Paine, "A Survey of the Boys' Technical High School of Milwaukee, Wisconsin, and the Organization of Its Automotive Department" (master's thesis, Iowa State College, 1928); Melvin S. Lewis, *Analysis of the Automechanic's Trade with Job Instruction Sheets,* Division of Vocational Education of the University of California and the State Board of Education, Trade and Industrial Series No. 4 (Berkeley: University of California, 1925); Oakland Public Schools, *Automobile Repair in the Vocational Continuation School,* Superintendent's Bulletin No. 33, Course of Study Series (Oakland, Calif.: Oakland Public Schools, 1922); Lewis S. Neeb, "The Automobile as a Subject of Instruction in the Public Secondary Schools" (master's thesis, University of Arizona, 1927); and Lynn C. McKee, "A Trade Training Curriculum in Automobile Mechanics for Senior High Schools" (master's thesis, Duke University, 1931).

18. Barley and Orr, *Between Craft and Science.*

19. Education researcher Mike Rose recently described the pervasive dichotomy between vocational and academic curricula as shaping "the kind of instruction one receives, the crowd one associates with, the status one feels in the yards and corridors." Rose criticized the mistaken belief that these two kinds of intelligence are distinct and intrinsically of different value in society. He suggests that we reconsider how we describe and attribute intelligence in school and in the workplace. Such an approach might significantly improve the system I muddled through. Mike Rose, *The Mind at Work: Valuing the Intelligence of the American Worker* (New York: Viking, 2004), 166.

One • The Problem with Chauffeur-Mechanics

1. "Chauffeurs Lord It over Their Employers," *New York Times,* 12 August 1906.

2. George Basalla, *The Evolution of Technology* (Cambridge: Cambridge University Press, 1988), 26–63; Thomas P. Hughes, *American Genesis: A Century of Invention and Technological Enthusiasm, 1870–1970* (New York: Penguin Books, 1989), 75–83; and D. O. Edge, "Technological Metaphor," in *Meaning and Control,* ed. D. O. Edge and J. N. Wolfe (London: Tavistock Publications, 1973), 31–59.

3. Americans of the period were, of course, familiar with railroad transportation, but

few were familiar with the management and care of railroads in the way that they were familiar with the management and care of horses and horse-drawn vehicles. For a census-based analysis of the development of horse-drawn transportation networks in New York State, and in particular the strong correlation between increasing personal wealth and increasing ownership of private vehicles during the nineteenth century, see Doris Halowitch, "A Ride into History: The Horse-Drawn Vehicle in Selected New York State Counties, 1800–1920," *Nineteenth-Century American Carriages: Their Manufacture, Decoration and Use* (Stony Brook, N.Y.: Museums at Stony Brook, 1987), 66–113.

4. In the interest of brevity "wealthy Americans" and "wealthy motorists" here include the upper-middle class.

5. Quote is from a gentleman's manual on equine care and equipment: Francis M. Ware, *Driving* (New York: Doubleday, Page, and f, 1903), 154. See also James Garland [Jorrocks, pseud.], *The Private Stable: Its Establishment, Management, and Appointments* (Boston: Little, Brown, Co., 1899), 325–26.

6. Quote is from Ware, *Driving,* 155–56. On commissions to domestic servants, see Daniel E. Sutherland, *Americans and Their Servants: Domestic Service in the United States from 1800 to 1920* (Baton Rouge: Louisiana State University Press, 1981), 68–69. G. P. Witson suggests that the practice of servant commissions and "tipping" goes back at least a century or more in Anglo-American history. *A Letter to the Gentlemen of Great Britain and Ireland on the Rate of Wages They Are Now Paying to Their Men-Servants* (London: T. and J. Allman, 1823).

7. Garland, *Private Stable,* 326.

8. Livery for coachman and groom ideally would be in the family's colors or in colors that complemented the carriage and equipage. Ware describes in some detail the prescribed livery for coachmen, grooms, and footmen in terms of tailoring, color, and material. Garland discusses and illustrates proper livery for stable servants in great detail. See Ware, *Driving,* 161–67; and Garland, *Private Stable,* 313–57.

9. Garland, *Private Stable,* 364–65.

10. For a useful typology of early automobiles, which unfortunately disregards electric and steam vehicles, see Peter J. Hugill, "Technology and Geography in the Emergence of the American Automobile Industry, 1895–1915," in *Roadside America: The Automobile in Design and Culture,* ed. Jan Jennings (Ames: Iowa State University Press, 1990), 29–39.

11. Dyke wrote, "In addition to the special tools usually supplied with a car, the following articles will be found of great use: a small pipe wrench, a pair of gas pipe pliers, a large and small screwdriver, a pair of flat-nosed pliers, a small hammer, a pair of wire cutters, a large jack knife, a flat, a half-round, and three-cornered file, a coil of soft iron, a roll of sticky tape, a cold chisel, a small ball pene [*sic*] hammer, a monkey wrench and some extra nuts and bolts, a few links of extra chain, including a match link, also extra chain for the pump, including mated link, a piece of asbestos for making gaskets, [t]ire repair tools, [a] good jack, [a] can of cylinder oil, [a] can of grease, [and] extra [spark] plugs." Andrew Lee Dyke and G. P. Dorris, *Diseases of the Gasoline Automobile and How to Cure Them* (St. Louis: A. L. Dyke Automobile Supply Co., 1903), 179. See also M. D. Blank, "Roadside Repairs and Expedients," *Horseless Age* 19, 6 March 1907, 330; "Emergency Kinks," *Horseless Age* 19, 20 March 1907, 423; and "Uses of Twine," *Horseless Age* 21, 15 January 1908, 63.

12. *Outing* 35, September 1900, 703.

13. Pierce Motor Car Co., *Instruction Book: Pierce Great Arrow Motor Car* (Buffalo, N.Y.: George N. Pierce Co., 1907); Morris Library, Special Collections, University of Delaware.

14. Electrics were, relatively speaking, maintenance free, but due to their limited power and range, they never factored into the chauffeur-driven, touring car market. This discussion of the maintenance requirements of early motor vehicles is based on W. B. Harsel, "The Chauffeur," *Horseless Age* 17, 14 February 1906, 275–76; "How to Grind Valves," *Motor World* 7, 15 October 1903, 103; X-Ray, "Reflections of a Repairman," *Horseless Age* 15, 10 May 1905, 529; J. A. Kingman, "The Care of the Automobile," *Outing Magazine* 38, July 1901, 433–36; and Albert L. Clough, "Care and Maintenance of Electric Vehicles," Frank S. Hanchett, "Care and Operation of Automobile Steam Boilers and Engines," and F. E. Watts, "Hints on the Care of a Gasoline Motor," all in *Horseless Age* 17, 30 May 1906, 759–68. See also David Kirsch, *The Electric Vehicle and the Burden of History* (New Brunswick, N.J.: Rutgers University Press, 2000); Gijs Mom, *The Electric Vehicle: Technology and Expectations in the Automobile Age* (Baltimore: Johns Hopkins University Press, 2004); and Michael Brian Schiffer, *Taking Charge: The Electric Automobile in America* (Washington, D.C.: Smithsonian Institution Press, 1994).

15. Irving Bacheller, *Keeping Up with Lizzy* (New York: Grosset and Dunlap, 1910), 16–17.

16. "Repairs Are Difficult," *Horseless Age* 8, 21 August 1901, 441.

17. "Cost of a Chauffeur," *Automobile Dealer and Repairer* 1, March 1906, 31; "Clothes for Master and Man," *Automobile Magazine* 4, March 1902, 234–35; "The Competent Chauffeur," *Automobile Dealer and Repairer* 1, April 1906, 71.

18. "Chauffeur Problem Solving Itself," *Motor World* 13, 28 June 1906, 165.

19. See "School for Chauffeurs: Locomobile's Experiment Panning Out Well," *Motor World* 9, 20 October 1904, 144; and "Where They Teach All about Autos," *New York Times*, 8 September 1907.

20. "Coachmen and Chauffeurs," *Automobile Dealer and Repairer* 1, July 1906, 172.

21. See Nicholas Papayanis, *The Coachmen of Nineteenth-Century Paris: Service Workers and Class Consciousness* (Baton Rouge: Louisiana State University Press, 1993), 191–206; "College for Chauffeurs: Practical Work Done at Locomobile Factory," *Motor World* 8, August 1904, 548; "School for Chauffeurs"; "Where They Teach All about Autos"; "Coachmen Becoming Chauffeurs," *Motor World* 20, 10 June 1909, 422; and "Chauffeur Problem Solving Itself."

22. The following discussion is based on data drawn from U.S. Bureau of the Census, *Thirteenth Census of the United States Taken in the Year 1910*, vol. 4: *Population, 1910: Occupation Statistics* (Washington, D.C.: GPO, 1914), 152–53, 181, 540, 545, 572, 575–78, 589.

23. Census enumerators were instructed to differentiate between chauffeurs who worked for commercial firms and those who worked for private individuals. That distinction was not made, however, in the published reports. An examination of the census manuscripts would enable one to disaggregate the numbers, but such an exercise would be questionable. Extensive reading of the available contemporary literature about chauffeurs indicates that there was probably very little difference in the urban-rural distribution, age, national origin, or race between private and commercial chauffeurs.

24. This lack of chauffeurs in southern cities can be attributed to the slightly later dif-

fusion of the automobile in the South. Atlanta hosted the first major automobile show in the region in 1909. For a case study of Atlanta, see Howard L. Preston, *Automobile Age Atlanta: The Making of a Southern Metropolis, 1900–1935* (Athens: University of Georgia Press, 1979).

25. Hostlers were those who took care of horses, mules, and their equipment, not including privately employed coachmen.

26. Short of combing through the U.S. Census manuscripts for each city, it is difficult to speculate about the age, nativity, and race profile of coachmen with any more specificity than the national figures. See U.S. Bureau of the Census, *Thirteenth Census of the United States*, vol. 4: *1910*, 414–15, 430–31.

27. "Confessions and Criticisms of a Chauffeur," *Motor World* 15, 10 January 1907, 111. Some chauffeurs also seem to have been paid well: from $70 to $130 per month for "machinist-operators" and one reported example of $6,000 per year. See "Two Kinds of Chauffeurs," *Motor World* 8, 2 June 1904, 359; "The Chauffeur and His Pay," *Motor World* 8, 2 June 1904, 370; and "Chauffeur's Pay $6000, Mr. Morris Tells Y.M.C.A.," *New York Times*, 2 November 1905.

28. "Chauffeurs vs. Coachmen," *Automobile Magazine* 3, September 1901, 809.

29. See "Chauffeurs Lord It over Their Employers"; "The Chauffeur's Duties," *Automobile Dealer and Repairer* 1, April 1906, 72; and "Chauffeurs Rebel," *Automobile Dealer and Repairer* 10, February 1911, 54. In the decades around the turn of the century, wealthy Americans who employed female domestic servants felt themselves plagued by a "servant problem" in general. The female domestic servant problem manifested superficial similarities with the chauffeur problem. At root, however, they differed in important ways. In a study of female domestic servants from 1870 to 1920 David Katzman found that the combination of an increase in the number of middle-class households that could afford domestics, the increasing emphasis on domestic cleanliness, and the relative decline in the number of women and girls choosing to go into domestic service resulted in the servant problem. The social isolation of live-in work and the degradation of servility caused many women to leave domestic service and their daughters not to enter, preferring instead newly available clerical and factory jobs. As Faye Dudden notes, women went into domestic work often as a last resort. In contrast, many young men appear to have been eager to enter chauffeur work. Chauffeuring represented the opportunity to use and to learn about the new technology of the automobile. It may also have been viewed by some as a stepping stone to other auto-related jobs. Most important, the servile status of the chauffeur's job was not firmly established during the first decade of the twentieth century, as was that of female domestic work. See David Katzman, *Seven Days a Week: Women and Domestic Service in Industrializing America* (Urbana: University of Illinois Press, 1978); Faye E. Dudden, *Serving Women: Household Service in Nineteenth-Century America* (Middletown, Conn.: Wesleyan University Press, 1983); Sutherland, *Americans and Their Servants;* and Pamela Horn, *The Rise and Fall of the Victorian Servant* (Wolfeboro Falls, N.H.: Alan Sutton Publishing Co., 1991).

30. The owners of the elegant Mount Washington Hotel in the White Mountains of New Hampshire renovated a nearby farmhouse—separate from the servants' quarters—to serve as a 100-room hotel for chauffeurs in 1907. That chauffeurs' hotel is now designated a National Historic Landmark and is open for guests year-round under the name "the

Bretton Arms." See Bryant Tolles, *The Grand Resort Hotels of the White Mountains: A Vanishing Architectural Legacy* (Boston: D. R. Godine, 1997), 227. See also "Chauffeur's Status: Hotelkeepers Wish to Know Where to Place Them during the Meal Hour," *Motor World* 6, 25 June 1903, 492; "Chauffeur on Chauffeur Status," *Motor World* 17, 13 February 1908, 868c–68d.

31. C. N. and A. M. Williamson, *The Lightning Conductor: The Strange Adventures of a Motor-Car* (New York: Henry Holt Co., 1903); quotes are from 21, 53–54, and 94, respectively.

32. As late as 1924, 43 percent of the new cars produced in the United States were open-bodied models, according to the National Automobile Chamber of Commerce (NACC), *Facts and Figures of the Automobile Industry, 1926* (New York: NACC, 1926), 8. The first three chapters of Paul C. Wilson, *Chrome Dreams: Automobile Styling since 1893* (Radnor, Pa.: Chilton Book Co., 1976), provide a good discussion of the design influences on the open-body automobile and the transition to closed-body design. Descriptions and floor plans of three different examples of private garages—one a converted horse stable—can be found in "The Private Garage," *Horseless Age* 20, 4 December 1907, 799–800. For a discussion of the architectural evolution of private garages, see Leslie G. Goat, "Housing the Horseless Carriage: America's Early Private Garages," in *Perspectives in Vernacular Architecture*, ed. Thomas Carter and Bernard L. Herman (Columbia: University of Missouri Press, 1989), 3:63–72.

33. James J. Flink, "Three Stages of American Automobile Consciousness," *American Quarterly* 24, October 1972, 456. The fear of fires associated with automobile fuels led to strict fire codes and regulations concerning the construction and location of automobile garages. Consequently, the Municipal Explosives Commission handled all garage ordinances in New York City during the first decade of the twentieth century. Elsewhere, municipalities and state courts constantly debated the question of how much gasoline could be safely stored, in what kind of containers, and in what type of buildings. Motorists for their part often used commercial garages due to the high insurance rates for storing a car in a private stable and "the impossibility of securing permission to keep about the premises an adequate supply of gasoline in existing private structures." See "The Growth of the Garage Business," *Horseless Age* 25, 26 April 1905, 477; "Storing Gasoline in Frame Garages," *Horseless Age* 17, 30 May 1906, 758; "New York Garage Regulations," *Horseless Age* 17, 20 June 1906, 938, and "Proposed New York Garage Regulations Bring Storm of Protest," *Horseless Age* 28, 19 July 1911, 103.

34. See "Typical American Garages: The New York Decauville Garage," *Horseless Age* 17, 30 May 1906, 781–83; and, "The Eureka Auto Station, New York," *Horseless Age* 23, 21 April 1909, 512–22.

35. See "Garages—Scarcity Noted in New York City," *New York Times*, 27 November 1905; "The Chauffeur; A Problem of the Day," *Motor World* 5, 12 March 1903, 905, 909–11; "Striking at Graft," *Motor World* 11, 28 December 1905, 689; "Confessions and Criticisms of a Chauffeur," *Motor World* 15, 10 January 1907, 111–12; "Plain Talk," *Garage* 1, December 1910, 5; R. E. Former, "The Lamentable Status of the Garage Business," *Horseless Age* 19, 6 February 1907, 201–2; X-Ray, "Comment on Trade Topics," *Horseless Age* 15, 3 May 1905, 508–9, and "Why Some Sales Are Not Made, II," *Horseless Age* 20, 7 August 1907, 172–73.

36. See "Dealers and Clubmen Meet: They Discuss Chauffeurs and Commissions, and

Bring Out Some Interesting Information," *Motor World* 9, 16 March 1905, 1214. For other mentions of the connection between coachmen's commissions and chauffeurs' commissions, see "The Chauffeur: A Problem of the Day," *Motor World* 5, 12 March 1903, 909; "Graft Evils in New York," *Automobile Dealer and Repairer* 1, April 1906, 54–55; and "Kicking Over the Old Man," *Motor Age* 23, 17 April 1913, 5–9.

37. See "The Chauffeur Problem," *Motor World* 9, 24 November 1904, 342; "Light on Chauffeur's Ways," *Motor World* 9, 8 December 1904, 427–28. For contemporary press accounts of accidents attributed to chauffeur joyriding, see "Use without Permission," *Automobile Dealer and Repairer* 1, May 1906, 112–13. "Drink-Maddened Chauffeur Cause of Auto Crash," *Press* (Philadelphia), 8 October 1904; "$10,000 Auto Wrecked: Chauffeur's Party Hurt," *New York Times*, 4 March 1905; "Bowery Crash at Dawn," *New York Times*, 18 June 1905; and "Joy Riders Wreck Toll Man's Home," *Philadelphia Inquirer*, 28 April 1909. Some newspaper editors seem to have been eager to publish lurid accounts of auto accidents not just for the obvious sensationalist reasons. Those favorable to motoring interests may also have published reports of reckless chauffeurs in an attempt to direct public and legislative attention away from the recklessness of wealthy motorists who drove their own vehicles. See, e.g., the *Philadelphia Inquirer* editorial from 10 April 1909 holding up auto owners as law-abiding citizens and reviling chauffeurs as reckless rogues who should be arrested and given "the full limit of the law." This editorial bias also permeates the other press accounts cited here.

38. The March 1909 report of the Law and Ordinance Committee of the Philadelphia Automobile Club credits committee chairman S. Boyer Davis, also chairman of the Legislative Committee of the Pennsylvania Motor Federation, with preparing the bill presented to the Pennsylvania Legislature by Representative Harrison Townsend which eventually became the motor vehicle law of 1909. The club's January 1910 newsletter boasts that "the new Pennsylvania State law was brought into being largely through the work of this club." For the auto club's account of the legislation, see Philadelphia Automobile Club, *Monthly Bulletin*, March 1909, 11; and "1910 Retrospect and Forecast," *Monthly Bulletin*, January 1910, 2. Quote is from *Laws of the General Assembly of the Commonwealth of Pennsylvania, Session of 1909*, act no. 174, sec. 23 (Harrisburg, Pa.: Harrisburg Publishing Co., State Printer, 1909), 272.

39. See *Lotz v. Hanlon*, 217 Pennsylvania Supreme Court 339 (1907); *Durham v. Struass*, 38 Pennsylvania Superior Court 620 (1909); and *Curran v. Lorch*, 243 Pennsylvania Supreme Court 247 (1913).

40. On the Kansas case, see "A Kansas Precedent," *Garage* 1, December 1910, 2–3. On the New York Case, see "Garage Responsible for Auto Wreck," *New York Times*, 27 December 1908; and "A Blow at 'Joy Rides,'" *New York Times*, 24 March 1909. See also the editorial debate between *Motor World* and *Horseless Age* over the merits of making garage owners liable for joyriding: "Another Blow at the Garageman," *Horseless Age* 28, 20 December 1911, 917; and "Not a Blow at the Honest Garageman," *Motor World* 30, 28 December 1911, 33.

41. Car owners, their friends, and relatives did not need to be licensed to drive in the state. This aspect of the 1909 act was definitively decided by Justice J. Staake in *Commonwealth v. Cooper*, 19 Pennsylvania District Reports 271 (1910).

42. *Laws of the General Assembly, 1909,* 267. The various legislative and legal actions taken by wealthy motorists reflect some of the broader reform strategies of the Progressive Era in which they appeared. In part, financially comfortable, urban-dwelling Progressive reformers compensated for the loss of older, face-to-face community and social ties by embracing regulation, bureaucratization, and professionalization. Some of the resulting coercive reforms of the era, such as temperance laws, immigration restriction, and southern Jim Crow laws, emerged from "the desire of native-born Americans to use social institutions and the law to restrain and direct the unruly masses, many of whom were foreign-born or black." Likewise, wealthy motorists worked through their statehouses and courtrooms to construct legal boundaries where older, social boundaries had failed. See Robert H. Wiebe, *The Search for Order, 1877–1920* (New York: Hill and Wang, 1967); and Arthur S. Link and Richard L. McCormick, *Progressivism* (Arlington Heights, Ill.: Harlan Davidson, 1983), 67–73, 96–104.

43. West Side Young Men's Christian Association (YMCA), *Automobile School for Owners, Prospective Owners and Chauffeurs,* brochure, 1904, in Vertical Files—Automobiles—Schools, Transportation Collection, National Museum of American History; hereafter cited as "Transp. Coll., NMAH."

44. A description of the Boston YMCA course offerings appears in "School for Motorists: Boston Y.M.C.A. Plans Elaborate Scheme of Lectures and Classes—The Details," *Motor World* 8, 29 October 1903, 170. A photograph of students working in the garage of "The Y.M.C.A. Chauffeur School" in New York appears in the Sunday pictorial section of the *New York Times,* 5 November 1905. See also "Y.M.C.A. Opens School for Drivers," *New York Times,* 10 November 1904; and "School of Automobile Instruction at West Side Y.M.C.A.," *New York Times,* 23 January 1910. Regarding the one thousand dollar gift from the Automobile Club of America, see Education Committee Minutes, West Side Branch, for 12 March 1904, at the YMCA of Greater New York Archives, New York; hereafter cited as "YMCA/NY."

45. See YMCA, *Automobile School* (1904); and a classified ad placed by the YMCA Employment Bureau in *Automobile Dealer and Repairer* 1, April 1906, 78. This involvement of the YMCA is, again, typical of the reform efforts of social progressives of the time. Progressive educators offered solutions "for almost every social problem of the early twentieth century." See "'Y' Chauffeurs Always Get Jobs . . . Student Drivers Feared as a Menace to the Policy of Time-Honored Graft," *New York Times,* 27 February 1907; and Link and McCormick, *Progressivism,* 90–92.

46. For comparative enrollment and revenue figures, see Education Committee Minutes, West Side YMCA, 10 December 1905 and 15 January, 7 May, and 5 November 1906. For a cumulative tally by year of the Auto School enrollment, see Education Committee Minutes, West Side YMCA, of 10 March 1916, YMCA/NY. Training chauffeurs and educating motorists about the new technology was a growth industry from mid-decade through World War I, and numerous private auto schools joined the YMCA in offering automobile courses. For a description of the courses offered at the New York School of Automobile Engineering, see "A Typical Institution for Training Automobile Drivers," *Horseless Age* 23, 13 January 1909, 39. On the Stewart Automobile School, see "School Turns Out Real Chauffeurs," *Automobile* 20, 17 June 1909, 1005. On the Portland (Maine) Auto School,

see school brochure, 1907, in Automobile Industry, box 20, folder 18, the Warshaw Collection of Business Americana, Automobile Industry, Archives Center, National Museum of American History; hereafter cited as "Warshaw Coll., NMAH." See also letters from the Boston Auto School and from C. A. Coey's School of Motoring, both in Automobile Industry, box 20, folder 18, Warshaw Coll., NMAH.

47. "How We Got a Chauffeur," *Automobile Dealer and Repairer* 1, August 1906, 217.

48. The anti-commission rhetoric of garage owners, while evident earlier, took on new clarity and stridency late in the course of the chauffeur problem. The heightened rhetoric closely followed the implementation of bureaucratic surveillance systems, described later, and therefore seems aimed at enrolling car owners' support for these systems. This late-period rhetoric provides some of the clearest formulations of garage owners' dissatisfaction with the chauffeur commission system and of their self-perception as being caught between unscrupulous chauffeurs and ignorant or unconcerned car owners. See "The Real Remedy," *Garage* 1, November 1910, 2; "Owner vs. Owner," *Garage* 1, December 1910, 2; "One Cause of 'Chauffeur Evils': Garageman Contends 'Aloofness' of the Car Owners Unfavorably Affects Conditions," *Motor World* 31, 9 May 1912, 26; and "Car Owners Can Correct the Graft Evil, Which Is Strong in the Garage Business," *Automobile Trade Journal* 18, July 1913, 163.

49. R. E. Former called for such a solution in "The Lamentable Status of the Garage Business," *Horseless Age* 19, 6 February 1907, 201–2.

50. "Garage Owners Organize: Plan General Reform in Management—Bad Chauffeurs to Go," *New York Times,* 20 April 1910; "Garage Owners Begin Business," *New York Times,* 8 May 1910; and "Garage Owners Organize: Charles D. Chase Appointed Manager of New Auto Association," *New York Times,* 25 May 1910.

51. A sense of the chronology of this increased interest in bureaucratic control within garages can be gained by comparing the 30 May 1906 garage issue of *Horseless Age,* which reprinted only three samples of such forms, with the 21 April 1909 garage issue, in which were reprinted dozens of examples of chauffeur tracking sheets, job cost forms, stockroom records, sales slips, and other forms.

52. J. Grant Cramer, "System of Checking Cars as to Arrival, Departure, and Who Uses Them," *Horseless Age* 17, 30 May 1906, 818.

53. See Philadelphia Automobile Club, *Monthly Bulletin,* December 1909, 12; and *Monthly Bulletin,* August 1910, 4.

54. "Typical American Garages: The New York Decauville Garage," *Horseless Age* 17, 30 May 1906, 781–83.

55. Joseph B. Baker, "Keeping Tab on Joy Riders," *Horseless Age* 25, 4 May 1910, 645–46.

56. Garages were urged to establish a system for tracking the movements of cars and chauffeurs as early as 1904, but such systems were apparently not widely adopted until about 1909. See "Crusade against Irresponsible Chauffeurs," *Horseless Age* 14, 28 September 1904, 327; and "Growth of the Garage Business," *Horseless Age* 23, 14 April 1909, 491.

57. Charles Zabriskie, New York City, to his son, 12 January 1910, rpt. in Robert D. Marcus and David Burner, eds., *America Firsthand,* vol. 2: *From Reconstruction to the Present,* 3rd ed. (New York: St. Martin's Press, 1995), 150. Historian James Flink wrote that "standard quick-demountable rims first came into general use in 1904." Yet Zabriskie's letter as

well as a text by Victor Pagé indicate that the non-demountable, or "clincher" type, wheels were likely still in wide use some nine years later. See Flink, *America Adopts the Automobile,* 285; Victor Pagé, *The Modern Gasoline Automobile: Its Design, Construction, and Repair* (New York: Norman W. Henley Publishing Co., 1913), 509–59; "Tire Troubles No Longer a Bugbear," *Horseless Age* 25, 16 February 1910, 258; and David Hebb, "No Punctures, No Blowouts," *Automobile Quarterly* 2 (Summer 1963): 222–31.

58. The low-tension "make and break" spark contacts and vibrating coil, or "buzz box," ignition systems of 1900 began to give way to spark plugs and high-tension magnetos by mid-decade, with automatic spark advance mechanisms being added by 1910. Ford, following the lead of French and English automakers, began using newly available vanadium steel alloys to produce more durable "gears, crankshafts, connecting rods, springs and drive shafts" for the 1908 Model T. The manufacturers of large, chauffeur-driven autos were also the first to offer self-starters to replace the hand crank, employing compressed air or acetylene gas and then electric motors after 1911. See Society of Automotive Engineers (SAE), *A History of the Automotive Internal Combustion Engine* (Warrendale, Pa.: SAE, 1976), 9–11; Thomas J. Misa, *A Nation of Steel: The Making of Modern America, 1865–1925* (Baltimore: Johns Hopkins University Press, 1995), 223–38; Clyde H. Pratt, *The Automobile Instructor* (Chicago: Shrewsbury Publishing Co., 1917), 53–96; Victor W. Pagé, *Automobile Starting, Lighting and Ignition Systems* (New York: Norman W. Henley Publishing Co., 1917); Ralph C. Epstein, *The Automobile Industry: Its Economic and Commercial Development* (Chicago: A. W. Shaw Co., 1928), 91–92, 157; C. E. Palmer, "Self-Starters for Automobile Engines," *Horseless Age* 28, 30 August 1911, 305–8; "Statistics of the 1913 Cars at the New York Show," *Horseless Age* 31, 29 January 1913, 246–47; T. A. Boyd, "The Self-Starter," *Technology and Culture* 9 (October 1968): 585–93; and Stuart W. Leslie, *Boss Kettering* (New York: Columbia University Press, 1983), 44–51.

59. The National Chauffeurs' Association was organized in October 1912 and began publishing a journal entitled the *Chauffeur.* For an overview of the organization's aims and objectives, see "The Chauffeurs' First Birthday! What We Started Out to Do and How We Are Doing It," *Chauffeur* 2, October 1913, 9–11, 32. See also the association's manifesto denying blacks membership, printed inside the back cover of the same issue. On outbreaks of racial tensions among New York chauffeurs, see "Chauffeurs Draw the Color Line," *Motor World* 20, 1 July 1909, 555; and "'Queering' the Negro Driver: The Process Lands Three Men in Jail," *Motor World* 22, 24 February 1910, 560a.

60. See "Chauffeurs Have a Bill," *New York Times,* 21 January 1910. I have not been able to find any records of the Chauffeurs' Professional Club of America with which to verify this newspaper account.

61. By the time of the 1920 census, the swelling ranks of the chauffeurs included relatively fewer foreign-born whites but a slightly increased proportion of blacks. See U.S. Bureau of the Census, *Fourteenth Census of the United States Taken in the Year 1920,* vol. 4: *Population, 1920, Occupations* (Washington, D.C.: GPO, 1923), 1053–1248.

62. Chauffeur employment data are from U.S. Bureau of the Census, *Fourteenth Census of the United States,* 39. The etymology of the word *chauffeur* parallels the social evolution of the position. Adopted from the French *chauffeur,* meaning "to heat or to stoke a fire," the French applied it to skilled automobilists in the late nineteenth century. It is

no surprise that Americans adopted the French term rather than the alternative English term *driver* because the best early automobiles were imported from France. Some wealthy American motorists even brought over expert French mechanics, whom they called "chauffeurs," to care for their new cars. Yet during the early 1900s American usage remained flexible, and *chauffeur* could mean a wealthy automobile enthusiast, such as Willie K. Vanderbilt; a privately employed driver-mechanic, such as the topic of this study; a commercially employed delivery or taxi driver, as in the Brotherhood of Teamsters, Chauffeurs, Warehousemen, and Helpers of America; or anyone who drove an automobile, whether for hire or pleasure, much as we would now say the "driver" of the car. As the chauffeur problem mushroomed in the 1900s and was eventually resolved by the mid-1910s, the term *chauffeur* stabilized in meaning as paid driver-mechanic and eventually as merely paid driver. On the new type of chauffeur as professional driver, see Charles B. Hayward, "The Status of the Chauffeur," *Horseless Age* 28, 6 December 1911, 846–47; and Keith Marvin, "The American Chauffeur: A Sociologic Appraisal," *Antique Automobile* 48, July–August 1984, 8–14.

63. "Sifting the Chauffeurs," *Motor World* 12, 19 April 1906, 633.

64. Hudson Motor Car Co., *Service Inspection Manual for Hudson Mechanics* (Detroit: Hudson Motor Car Co., 1917), 6, Vertical Files—Automobiles—Hudson, Transp. Coll., NMAH.

Two • Ad Hoc Mechanics

1. Floyd Clymer, *Floyd Clymer's Historical Scrapbook: Early Advertising Art* (New York: Bonanza Books, 1955), 9.

2. Bellamy Partridge, *Fill 'er Up! The Story of Fifty Years of Motoring* (New York: McGraw-Hill, 1952), 153–54.

3. See David A. Hounshell, *From the American System to Mass Production, 1800–1932: The Development of Manufacturing Technology in the United States* (Baltimore: Johns Hopkins University Press, 1984), 189–215.

4. See Young and Co. advertisement in *Dana Burk's Riverside City Directory* (Riverside, Calif.: Riverside Directory Co., 1907), 299.

5. See "Sydney B. Bowman Cycle Co.," brochure, 1901; and "Sydney B. Bowman Automobile Co.," brochure, 1902, both at Henry Ford Museum and Greenfield Village Research Center, Dearborn, Mich.; hereafter cited as "HFMGV."

6. "Making Ready for Motorcycles," *Bicycling World* 42, 18 October 1900, 48.

7. "Rudiments of the Motocycle [*sic*]" *Bicycling World* 42, 10 January 1901, 374–75. Over the next ten years the journal's masthead reflected the shifting editorial focus as the words *Bicycling World* grew smaller, and *Motorcycle Review* grew larger.

8. "Why Thousands Call Carl" brochure, ca. 1932–33, Call Carl Files, Promotion Folder, Transp. Coll., NMAH.

9. "Repair of Automobiles," *Blacksmith and Wheelwright* 66, December 1912, 454.

10. See Joseph J. Corn, "Work and Vehicles: A Comment and Note," in *The Car and the City: The Automobile, the Built Environment, and Daily Urban Life*, ed. Martin Wachs and Margaret Crawford (Ann Arbor: University of Michigan Press, 1992), 25–34.

11. Letter from Oscar Friedrich, *Blacksmith and Wheelwright* 64, July 1911, 746.

12. "Answers to Correspondents," *Blacksmith and Wheelwright* 62, July 1910, 264.

13. Automotive articles in *Scientific American,* though not a trade journal per se, may have been another source of information for ad hoc mechanics.

14. Andrew Lee Dyke and G. P. Dorris, *Diseases of the Gasoline Automobile and How to Cure Them* (St. Louis: A. L. Dyke Automobile Supply Co., 1903). Dyke followed this book with *The Anatomy of the Automobile* (St. Louis: A. L. Dyke, 1904); *Dyke's Troubles, Remedies and Repairs of the Gasoline Engine* (St. Louis: A. L. Dyke, 1909); and his very popular *Dyke's Automobile and Gasoline Engine Encyclopedia* (St. Louis: A. L. Dyke, 1910); not to mention the automobile correspondence course he began offering in 1915. Leonard Elliott Brookes, *The Automobile Hand-Book: A work of Practical Information for the Use of Owners, Operators and Automobile Mechanics* (Chicago: F. J. Drake and Co., 1905); Charles P. Root, *Automobile Troubles and How to Remedy Them,* 5th rev. ed. (Chicago: Charles C. Thompson Co., 1909); Victor W. Pagé, *The Modern Gasoline Automobile* (New York: Norman W. Henley Publishing Co., 1912).

15. John Vander Voort, Daybooks, vol. 3: 22 October 1906–26 December 1916; and vol. 4: 27 December 1916–9 March 1923, MC 31, Special Collections and Archives, Rutgers University Libraries, New Brunswick, N.J.; hereafter cited as "Vander Voort Daybooks."

16. Robert E. Ireland, writing about early motoring in North Carolina, recounts as "perhaps typical" the story of John Adam Young, who began as a blacksmith but eventually took on automobile work: "His garage became known as Davidson Motor Co., and he began a forty-three-year association with Gulf Oil Corporation." *Entering the Auto Age: The Early Automobile in North Carolina, 1900–1930* (Raleigh: Division of Archives and History, North Carolina Department of Cultural Resources, 1990), 110. Henry Dominguez published numerous photographs of early Ford agencies, a couple of which were blacksmith shops, in *The Ford Agency: A Pictorial History* (Osceola, Wis.: Motorbooks International, 1980). Furthermore, the linkage between blacksmiths and auto mechanics was distilled, idealized, and reinforced in the public mind by images such as a widely published 1931 Ford Motor Co. advertisement featuring a four-color illustration of a blacksmith working at his forge. The ad copy read, "When you were younger, you learned that when Ford sold a car to a customer he followed up the sale by going to the best mechanic in town . . . the blacksmith, bicycle-repairman, or plumber . . . *and giving him a complete lesson in the maintenance and servicing of automobiles!*" This Ford advertisement appeared in *American Hebrew, American Magazine, Business Week, Fortune, Nation's Business, Popular Science Monthly, Review of Reviews, Time,* and *World's Work* during November and December 1931. Accession 19, box 115, Dealer Services—Service—1931, HFMGV. Kenneth Dunshee was less sanguine about the ability of most blacksmiths to make the transition to automobile work: "for most of the craftsmen [the automobile] meant the certain road to professional oblivion." *The Village Blacksmith: A Story of His Metal and Mettle* (Watkins Glen, N.Y.: Century House, 1957), 34.

17. Corn, "Work and Vehicles," 31.

18. Stephen L. McIntyre, "'The Repair Man Will Gyp You': Mechanics, Managers, and Customers in the Automobile Repair Industry, 1896–1940" (Ph.D. diss., University of Missouri, Columbia, 1995), 479–80.

19. The U.S. Census reported a small number of female blacksmiths in 1890, 1900,

and 1910, but the overwhelming majority of blacksmiths, wheelwrights, and carriage makers were male during the period under study. See U.S. Department of the Interior, Census Office, *Report on Population of the United States at the Eleventh Census: 1890* (Washington, D.C.: GPO, 1897), pt. 2, 304; U.S. Department of the Interior, Census Office, *Twelfth Census of the United States Taken in the Year 1900*, vol. 2: *Population* (Washington, D.C.: GPO, 1902), pt. 2, cxlvii; and *Thirteenth Census of the United States*, vol. 4: *1910*, 91.

20. Throughout the 1880s and 1890s the *American Machinist* heralded itself as "a journal for machinists, engineers, founders, boilermakers, patternmakers and blacksmiths." *American Machinist* dropped blacksmiths and all other trades from its masthead in 1900 to become "a practical journal of machine construction."

21. On the variety of new machinery employed by American farmers, see Douglas R. Hurt, "Agricultural Technology in the Twentieth Century," *Journal of the West* 30 (April 1991): 3–100; Reynold M. Wik, *Steam Power on the American Farm* (Philadelphia: University of Pennsylvania Press, 1953); and Roy Burton Gray, *Development of the Agricultural Tractor in the United States* (St. Joseph, Mich.: American Society of Agricultural Engineers, 1956).

22. Letter printed in *Dakota Farmer*, August 1889, qtd. in Wik, *Steam Power on the American Farm*, 234–35 n. 93.

23. On the dramatic increase in steam power on American farms during the decades around the turn of the twentieth century and its geographic distribution, see Wik, *Steam Power*, 82–107, 155–87. Although the steam behemoths of which Wik writes were mostly associated with the Midwest and Pacific, John Vander Voort was called on to repair the wheel of a steam traction engine in rural New Jersey in 1910. Vander Voort daybooks, vol. 3, entry for 21 February 1910. U.S. Census analysts placed the overall value of implements on American farms at $749.8 million in 1910, with surprisingly even distribution among the various states. See U.S. Department of the Interior, Census Office, *Twelfth Census of the United States Taken in the Year 1900*, vol. 10: *Manufactures* (Washington, D.C.: GPO, 1902), pt. 4, 356, table 9.

24. A sense of the geographic context of Vander Voort's community can be gained by examining F. W. Beers, *Atlas of Hunterdon County, New Jersey* (New York: Beers, Comstock, and Cline, 1873); and *Farm and Business Directory of Hunterdon and Somerset Counties, New Jersey* (Philadelphia: W. Atkinson, 1914), both in Special Collections and Archives, Rutgers University Libraries. On the social and technological changes affecting blacksmiths in rural Pennsylvania, see Jeannette Lasansky, *To Draw, Upset, and Weld: The Work of the Pennsylvania Rural Blacksmith, 1742–1935* (Lewisburg, Pa.: Oral Traditions Project, Union County Historical Society, 1980).

25. Circulation numbers for this journal are not available, but with a literacy rate of 94.1 percent in 1890 blacksmiths were likely familiar with it or its competitors. The overall literacy rate for the United States in 1890 was 86.66 percent. U.S. Census Bureau, *Eleventh Census: 1890*, xxx–xxxi, 770. For a short biography of Richardson and a listing of his other publishing endeavors, which included *Housekeepers' Companion, Idle Hours, Saw Mill Gazette, Shoe and Leather Manufacturer*, and eventually *Automobile Dealer and Repairer* beginning in 1906, see "Founder and for Thirty-Five Years Publisher of *The Blacksmith and Wheelwright*," *Blacksmith and Wheelwright* 71, January 1915, 479.

26. "Tools Needed in a Small Shop," *Blacksmith and Wheelwright* 37, March 1898, 36; "Tools Needed in a Small Shop—No. 2," *Blacksmith and Wheelwright* 37, April 1898, 56. See

also the 1887 inventory Benjamin Sebastian, Berks County, Pa., reproduced in Lasansky, *To Draw, Upset, and Weld,* 9.

27. "His Scale of Prices," *Blacksmith and Wheelwright* 35, March 1897, 238; "Repair Prices," *Blacksmith and Wheelwright* 40, September 1899, 444; "Prevailing Prices for Repairs in Different Places," *Blacksmith and Wheelwright* 40, October 1899, 474–75.

28. "Modern Blacksmith and Wheelwright Shop—I" and "Modern Blacksmith and Wheelwright Shop—II," *Blacksmith and Wheelwright* 35, March 1897, 238–39. George Hewes's shop in Lancaster County, Pa., used a less modern waterwheel to power a trip-hammer and a grindstone in a similar manner. See Lasansky, *To Draw, Upset, and Weld,* 8.

29. The Department of Agriculture's estimate of the number of farm horses continued to increase to 21.6 million in 1918, before beginning a steady decline. See U.S. Department of Agriculture, *Yearbook of the Department of Agriculture, 1914* (Washington, D.C.: GPO, 1915), 618, table 143; and *Yearbook of the Department of Agriculture, 1925* (Washington, D.C.: GPO, 1926), 1200, table 612.

30. "Horseless Carriages and Blacksmiths," *Blacksmith and Wheelwright* 41, January 1900, 39.

31. "Repairing Automobiles," *Blacksmith and Wheelwright* 44, September 1901, 750.

32. For particularly illustrative examples of advertisements, see those for High Wheel Auto Parts Co., Muncie, Ind.; and Cray Brothers, "Jobbers of Motor Car Supplies and Carriage Hardware," Cleveland, Ohio, in *Blacksmith and Wheelwright* 57, June 1908, 237; and 60, November 1909, 921, respectively.

33. "A Business Opportunity," *Blacksmith and Wheelwright* 50, November 1904, 474.

34. "Get Ready for Business," *Blacksmith and Wheelwright* 56, September 1907, 934.

35. "Automobile Repairing," *Blacksmith and Wheelwright* 45, June 1902, 252. See also "Automobile Repairs: How They Are Being Forced upon Blacksmiths and Wheelwrights," *Blacksmith and Wheelwright* 51, March 1905, 671.

36. "Automobile Repairing," *Blacksmith and Wheelwright* 47, March 1903, 670.

37. "The Village Rubbersmith," *Blacksmith and Wheelwright* 44, September 1901, 248. Another variation on Longfellow's verse appeared in 1916 extolling the riches of the "new village blacksmith," who "merely tinkers cars that pass along his way." "The New Village Blacksmith," *Blacksmith and Wheelwright* 74, October 1916, 363.

38. "She Repairs Autos: Although Miss Jones Is Rich, a Suffragette and a Public Speaker," *Blacksmith and Wheelwright* 72, October 1915, 833.

39. "New Prize Topic," *Blacksmith and Wheelwright* 50, December 1904, 511.

40. See editorials, "The Automobile Here to Stay," *Blacksmith and Wheelwright* 60, November 1909, 908; and "The Automobile in Business," *Blacksmith and Wheelwright* 61, June 1910, 224.

41. "Does No Auto Work," *Blacksmith and Wheelwright* 61, June 1910, 227.

42. See letter from Nick Jacobs, *Blacksmith and Wheelwright* 62, August 1910, 297.

43. "Automobile Work and Farm Work," *Blacksmith and Wheelwright* 65, February 1912, 68.

44. Rural animosity toward urban elite motorists is borne out by Wik's survey of period farm journals. See Reynold M. Wik, *Henry Ford and Grass-roots America* (Ann Arbor: University of Michigan Press, 1972), 15.

45. "Automobile Work and Farm Work," 68.

46. The latter contest was to run from August 1915 to January 1916 but was extended to May 1916 due to the "meager" response. Winners were announced in the November 1916 issue. See "A Prize Topic," *Blacksmith and Wheelwright* 72, August 1915, 755; "The Prize Topic," *Blacksmith and Wheelwright* 73, February 1916, 60; and "Prize Topic Awards," *Blacksmith and Wheelwright* 74, November 1916, 403.

47. *Blacksmith and Wheelwright* began a series of articles on "automobile welding" with oxy-acetylene outfits in May 1916.

48. "Among the Michigan Shops," *Blacksmith and Wheelwright* 77, January 1918, 23. Whether Hobart was faithfully reporting an actual conversation or had fabricated an idealized account for rhetorical purposes, evidently he encountered blacksmiths doing contract work for automobile repair shops with enough frequency that he felt the need to return to the issue again in March 1919. See "In a Clay County Smith Shop: An Article about a Blacksmith Who Didn't Want to Do Automobile Work Because It Interfered with His Regular Business," *Blacksmith and Wheelwright* 79, March 1919, 11.

49. On the ways rural Americans used automobiles to reshape important patterns in their social life such as schooling, churchgoing, health care, visiting, and shopping, see Wik, *Henry Ford and Grass-roots America;* Michael L. Berger, *The Devil Wagon in God's Country: The Automobile and Social Change in Rural America, 1893–1929* (Hamden, Conn.: Archon Books, 1979); Joseph Interrante, "You Can't Go to Town in a Bathtub: Automobile Movement and the Reorganization of Rural American Space, 1900–1930," *Radical History Review* 21, Fall 1979, 151–68; and Norman T. Moline, *Mobility and the Small Town, 1900–1930: Transportation Change in Oregon, Illinois,* Department of Geography Research Paper no. 132 (Chicago: University of Chicago Press, 1971).

50. Interestingly, queries published in the "Truck and Tractor Troubles" department of *Blacksmith and Wheelwright* seemed to be as much about troubles that readers were having with their own Fords as they were about troubles with customers' machines.

51. Ron Kline and Trevor Pinch, "Users as Agents of Technological Change: The Social Construction of the Automobile in the Rural United States," *Technology and Culture* 37, October 1996, 778–80.

52. Ford Motor Co., *Instruction Book for Ford Model T Cars,* 3rd ed. (1913; rpt., Detroit: Ford Motor Co., 1954).

53. Wik, *Henry Ford and Grass-roots America,* 35. Thomas S. Dicke cites more generous numbers: seven thousand Ford dealers in 1913, increasing to ninety-five hundred dealers and eighty-five hundred subdealers in 1924, and forty-five thousand authorized Ford service dealers in 1928. See Dicke, *Franchising in America: The Development of a Business Method, 1840–1980* (Chapel Hill: University of North Carolina Press, 1992), 72.

54. *Riverside City and County Directory, 1897–98* (Riverside, Calif.: A. M. Bushnell and Co., 1897); *Dana Burk's Riverside City Directory, 1907* (Riverside, Calif.: Riverside Directory Co., 1907); *Riverside City and County Directory, 1914* (Los Angeles: Riverside Directory Co., 1913); *Riverside City and County Directory, 1915* (Los Angeles: Riverside Directory Co., 1914); *Riverside City and County Directory, 1917* (Los Angeles: Riverside Directory Co., 1917); *Riverside City Directory, 1921* (Los Angeles: Riverside Directory Co, 1921). I make no claim that Riverside's was the "average" or "typical" experience for Americans of the period. No single community could shoulder such a burden, and a sampling of communities varied enough

in geographic and demographic characteristics to do so would be far too large an undertaking for the current project. Rather, this study represents a random sample, chosen on the basis of materials available to the author at the time of writing. With that said, it is doubtful whether the transition rate from blacksmithing to auto repair was significantly higher in other regions of the country, though further study is warranted.

55. See n. 54. Also, J. N. Russell and Son, general blacksmiths in Barry, Tex., wrote to *Blacksmith and Wheelwright* in 1910 describing their aggressive approach to attracting automobile business into their shop. One has to wonder how much the *Son* referred to in the shop's name had to do with this attitude. Letter from J. N. Russell and Son, *Blacksmith and Wheelwright* 61, June 1910, 227.

56. Aldren A. Watson, *The Village Blacksmith* (New York: Thomas Y. Crowell Co., 1968), 103–5.

57. Franklin K. Mathiews, "Blowing Out the Boy's Brains," *Outlook* 108, November 1914, 652–53; cited in Deidre Johnson, *Edward Stratemeyer and the Stratemeyer Syndicate* (New York: Twayne Publishers, 1993), 163.

58. Information about Edward Stratemeyer is culled from Johnson, *Stratemeyer Syndicate;* Diedre Johnson, *Stratemeyer Pseudonyms and Series Books: An Annotated Checklist of Stratemeyer and Stratemeyer Syndicate Publications* (Westport, Conn.: Greenwood Press, 1982); Carol Billman, *The Secret of the Stratemeyer Syndicate: Nancy Drew, the Hardy Boys, and the Million Dollar Fiction Factory* (New York: Ungar Co., 1986); Roger Garis, *My Father Was Uncle Wiggily* (New York: McGraw-Hill, 1966). For nineteenth-century antecedents to the series book, see Michael Denning, *Mechanic Accents: Dime Novels and Working-Class Culture in America* (London: Verso, 1987).

59. Other works that employed automobile adventure themes include Donald Grayson, *On High Gear, or, The Motor Boys on Top* (New York: Street and Smith, 1909); Roy Rockwood, *The Speedwell Boys in Their Racing Auto* (New York: Cupples and Leon, 1913); Victor Appleton, *Tom Swift and His Motorcycle, or, Fun and Adventure on the Road* (New York: Grosset and Dunlap, 1910); and *Tom Swift and His Electric Runabaout, or, The Speediest Car on the Road* (New York: Grosset and Dunlap, 1910). See also David K. Vaughn, "The Automobile in American Juvenile Series Fiction, 1900–1940," in *Roadside America: The Automobile in Design and Culture,* ed. Jan Jennings (Ames: Iowa State University Press, 1990), 74–81; and Clay McShane, *Down the Asphalt Path: The Automobile and the American City,* Columbia History of Urban Life, ed. Kenneth T. Jackson (New York: Columbia University Press, 1994), 144–48.

60. Marion Jennings, "Growing Up with the Automobile," interview by Rose D. Workman, Charleston, S.C., 10 February 1939, in "American Life Histories: Manuscripts from the Federal Writers' Project, 1936–40," Works Progress Administration (WPA) Federal Writers' Project Collection, Manuscript Division, Library of Congress, http://memory.loc.gov/ammem/wpaintro/wpahome.html.

Three • Creating New Mechanics

1. *Learn Autos Where Autos are Made: Y.M.C.A. Automotive School,* brochure, ca. 1918–19, Vertical File—Schools, Automotive History Collection, Detroit Public Library.

2. On the origin of the West Side YMCA auto school in the context of the chauffeur problem, see chap. 1.

3. On the decline of traditional apprenticeships in America, see W. J. Rorabaugh, *The Craft Apprentice: From Franklin to the Machine Age in America* (New York: Oxford University Press, 1986); Paul H. Douglas, *American Apprenticeship and Industrial Education,* Columbia University Studies in the Social Sciences (New York: Columbia University Press, 1921; rpt., New York: AMS Press, 1968), 11–84; Paul E. Johnson, *A Shopkeeper's Millennium: Society and Revivals in Rochester, New York, 1815–1837* (New York: Hill and Wang, 1978); and Nina Lerman, "From 'Useful Knowledge' to 'Habits of Industry': Gender, Race, and Class in Nineteenth-Century Technical Education" (Ph.D. diss., University of Pennsylvania, 1993).

4. See Ileen A. DeVault, *Sons and Daughters of Labor: Class and Clerical Work in Turn-of-the-Century Pittsburgh* (Ithaca, N.Y.: Cornell University Press, 1990); and Lerman, "From 'Useful Knowledge.'"

5. As early as 1866, sixty students preparing for the ministry studied Greek and Latin in YMCA classes offered by four United States chapters. William Orr, "Educational Work of the Young Men's Christian Associations," in U.S. Department of the Interior, Bureau of Education, *Biennial Survey of Education, 1916–1918,* vol. 1, bulletin, 1919, no. 88 (Washington, D.C.: GPO, 1921), 605.

6. William F. Hirsch, "Educational Work of the Young Men's Christian Association," in U.S. Department of the Interior, Bureau of Education, *Biennial Survey of Education, 1920–22,* vol. 1, bulletin, 1924, no. 13 (Washington, D.C.: GPO, 1924), 693.

7. Education Committee Minutes, West Side YMCA, 7 June 1902, YMCA of Greater New York Archives, New York; hereafter cited as "YMCA/NY."

8. The original study is unavailable but is cited in a letter from Harrison S. Colburn, educational director of the West Side YMCA, to the Educational Committee of the West Side YMCA, 19 December 1905. Letter is pasted into Education Committee Minutes, West Side YMCA, for December 1905.

9. Colburn letter, 19 December 1905. See also Terry Donoghue, *An Event on Mercer Street: A Brief History of the YMCA of the City of New York* (privately printed, n.d. [ca. 1951]), 58–60.

10. West Side YMCA, *Automobile School for Owners, Prospective Owners and Chauffeurs,* brochure, 1904, 2, in Vertical Files—Automobiles—Schools, Transportation Coll., NMAH; West Side YMCA, *Automobile School for Owners, Prospective Owners and Chauffeurs,* brochure, 1905–6, 2, box 122, Branches / West Side, Printed Material, 1897–1923, YMCA/NY.

11. Cumulative enrollment numbers for the first eleven seasons of the auto school are found in the Education Committee Minutes, West Side YMCA, 10 March 1916. Later enrollment numbers are scattered throughout the Education Committee minutes. Figures for 1 August 1916 through 1 August 1917 are found in the minutes of 11 October 1918, YMCA/NY.

12. *Automobile School* brochure, 1904; *Automobile School* brochure, 1905–6.

13. See, e.g., R.D.B., "Fake Automobile Schools," letter to the editor, *Horseless Age* 17, 28 March 1906, 484; and letter from Forrest R. Jones defending the similarly short curriculum of the Manhattan Automobile School, "Automobile Schools," *Horseless Age* 17, 25 April 1906.

14. One satisfied graduate of this class was Dr. Drenden, a New York City physician, who reported being "very well satisfied" with the course he took at the YMCA. "Interviews with Physicians and Other Users in Various Cities," *Horseless Age* 22, 23 September 1908, 407.

15. Education Committee Minutes, West Side YMCA, 10 March 1916, YMCA/NY.

16. On the character-building program at the West Side, see Education Committee Minutes, West Side YMCA, 30 October 1913, 11 February 1916, and 12 December 1921, YMCA/NY. By the 1920s the Christian character–building emphasis of the YMCA industrial and educational programs was explicit and emphatic in the literature generated by the national organization. Quote is from Charles R. Towson (member of the International Committee), "Industrial Program of the Young Men's Christian Association," *Annals of the American Academy of Political and Social Science* 103, September 1922, 134–37. See also Hirsch, "Educational Work"; Orr, "Educational Work"; and "Automotive School Opportunities," *Educational Messenger,* 15 March 1921, 3. *Educational Messenger* was the newsletter of the Educational Council of the YMCA International Committee. It is available at the YMCA of the USA Archives, University of Minnesota Libraries, Minneapolis.

17. Sociologist Paul Gilroy has speculated that African Americans' "histories of confinement and coerced labour must have given them additional receptivity to the pleasures of auto-autonomy as a means of escape, transcendence and perhaps even resistance." "Driving While Black," in *Car Cultures,* ed. Daniel Miller (Oxford: Berg, 2001), 84.

18. "Report of the Fourth Annual Convention of the National Negro Business League," Wilberforce, Ohio, 1903, 24; cited in August Meier and Elliott Rudwick, "The Boycott Movement against Jim Crow Streetcars in the South, 1900–1906," *Journal of American History* 55, March 1969, 756–75.

19. Boyd letter in Nashville *Banner* newspaper, 27 September 1905; cited in Meier and Rudwick, "Boycott Movement against Jim Crow Streetcars."

20. "Race Union in Nashville," *Cleveland Journal,* 13 January 1906; John Ingham and Lynne B. Feldman, "Richard Henry Boyd," *African American Business Leaders: A Biographical Dictionary* (Westport, Conn.: Greenwood Press, 1994), 106.

21. Blaine A. Brownell, "A Symbol of Modernity: Attitudes toward the Automobile in Southern Cities in the 1920s," *American Quarterly* 24 (March 1972): 20–44.

22. Kathleen Franz, " 'The Open Road': Automobility and Racial Uplift in the Interwar Years," in *Technology and the African-American Experience: Needs and Opportunities for Study,* ed. Bruce Sinclair (Cambridge, Mass.: MIT Press, 2004), 131–53.

23. Education Committee Minutes, West Side YMCA, 4 November 1908, YMCA/NY. This action supports Susan Kerr Chandler's observation that before 1910 YMCA policies excluding blacks were tacitly understood but potentially flexible. The Education Committee at least considered the proposal. Furthermore, its response—to set up segregated classes if enrollment warranted—foreshadowed the rigid physical segregation enforced by YMCAs throughout the North over the next generation, "Jim Crow, northern-style." After 1910, writes Chandler, "segregation [in the YMCA] was fully institutionalized . . . and remained intact until 1946." Chandler observes that the YMCA developed explicit policies of segregation in response to the twin "transforming events" of the Great Migration and millionaire Julius Rosenwald's gift to help build new, segregated YMCA buildings in twenty-five cities for black men and boys. Susan Kerr Chandler, " 'Almost a Partnership': African

Americans, Segregation, and the Young Men's Christian Association," *Journal of Sociology and Social Welfare* 21, March 1994, 97–111.

24. See report of Brokaw's visit to the Broadway Auto School, Education Committee Minutes, West Side YMCA, 11 February 1916, YMCA/NY. The national YMCA reported the opening of two new YMCA auto schools in Knoxville, Tenn., in 1920: "one for white and one for colored students." See "Automotive and Technical Schools," *Educational Messenger,* 15 October 1920, 3.

25. Pollard placed half-page advertisements in the first five issues of W.E.B. DuBois's important new black journal, the *Crisis,* in 1910–11. The ads included a photograph of two black men in clean white shirts and dark pants working on an automobile engine and listed the school address as the Hotel Maceo, 213 West 53 Street. The same address was given for the Broadway Auto School and Sales Co. in a 1915 *New York Age* article about black proprietor Ben Thomas. Broadway was reported to have graduated "more than fifteen hundred chauffeurs, men and women, white and colored," since its founding seven years earlier. It is not yet clear whether Cosmopolitan changed its name to Broadway or if the two schools operated separately. See *Crisis* 1, no. 1 (November 1910): advertisement, 18 f.; "Ben Thomas Invents an Auto Enclosure," *New York Age,* 16 September 1915, 1; and Franz, "Open Road," 140.

26. Virginia Scharff, *Taking the Wheel: Women and the Coming of the Motor Age* (New York: Free Press, 1991).

27. Education Committee Minutes, West Side YMCA, 30 October 1913, 6 April 1917, YMCA/NY. On the struggle between the YMCA and the YWCA over the general provision and delivery of services to women, see Jodi Vandenberg-Daves, "The Manly Pursuit of a Partnership between the Sexes: The Debate over YMCA Programs for Women and Girls, 1914–1933," *Journal of American History* 78, March 1992, 1324–46.

28. West Side YMCA, *Automobile Instruction,* brochure, 1909–10, 9, Automobile Industry, box 20, folder 18, Warshaw Collection, NMAH; Education Committee Minutes, West Side YMCA, 13 December 1920, 14 March 1921, YMCA/NY.

29. Education Committee Minutes, West Side YMCA, 18 March 1912, YMCA/NY.

30. Education Committee Minutes, West Side YMCA, 29 September and 8 December 1916, YMCA/NY.

31. H. C. Brokaw is first mentioned in the Education Committee Minutes, 7 December 1906. His promotion to director of Technical Schools was approved on 11 October 1918, YMCA/NY.

32. The education committee also considered printing their own textbook to sell to other YMCA auto schools, see Education Committee Minutes, West Side YMCA, 11 May 1917, YMCA/NY. Publication of Brokaw's book is noted in the minutes of 14 February 1910 and 11 May 1917. I presume both references were to H. Clifford Brokaw and Charles A. Starr, *Putnam's Automobile Handbook: The Care and Management of the Modern Motor-Car* (New York: G. P. Putnam's Sons), but the earliest extant edition I could locate was from 1918.

33. Education Committee Minutes, West Side YMCA, 11 June 1919, 11 October 1920, YMCA/NY.

34. Education Committee Minutes, West Side YMCA, 9 May 1919, YMCA/NY.

35. Education Committee Minutes, West Side YMCA, 12 December 1921, YMCA/NY; "Auto Schools—Advance the Spark," *Educational Messenger,* 3 May 1920, 2.

36. "Automotive and Technical Schools," *Educational Messenger,* 15 October 1920, 3.

37. Some automakers, such as Locomobile, offered training at their factory for customers or their "men," and the Automobile Mechanics Union of San Francisco reportedly opened a school for apprentices and journeymen in the spring of 1920. The largest private auto schools, however, were not directly affiliated with industry or labor. "School for Chauffeurs: Locomobile's Experiment Panning Out Well," *Motor World* 9, 1904, 144; "Auto Mechanics to Open Training School" *San Francisco Chronicle,* 7 March 1920.

38. Letter, C. A. Coey to C. Bullard, n.d. (ca. 1910), Automobile Industry, box 20, folder 18,Warshaw Collection, NMAH. This appears to be a follow-up letter to a prospective student. Coey's school letterhead bragged of being the only auto school "engaged in the automobile business" but that seems to have had little positive impact on the correspondence curriculum, which I had the opportunity to examine at an automobile literature swap meet but did not purchase.

39. *The Practical Auto School,* brochure, 1912, Vertical File—schools, Automotive History Collection, Detroit Public Library, Detroit, Mich.

40. The Sweeney Auto School moved into a new ten-story, million-dollar building in 1917 and boasted of having trained ten thousand men since opening for business in 1908. The Education Committee of the West Side YMCA was also conscious of the rapid growth of the Sweeney, Rahe, and Detroit schools. See *The $1,000,000 Sweeney Automobile and Tractor School,* brochure, 1917 Trade Catalog Collection, NMAH; Michigan State Auto School advertisement, *Motor World* 49, 4 October 1916, 95; "Better Mechanics for the Industry: How the Michigan State Auto School Is Training Men" *Motor Age* 40, 21 July 1921, 12–13; George H. Hawes, "Michigan State Automobile School of Detroit," *Automobile Dealer and Repairer* 36, October 1923, 17–20; Education Committee Minutes, West Side YMCA, 9 March 1917, 9 January 1920. On the Rahe Auto and Tractor School, see John Gunnell, "Early Auto Repair Schools," *California Highway Patrolman,* January 1995, 36–42.

41. "Automotive and Technical Schools," *Educational Messenger,* 15 October 1920, 3.

42. Quote is from "Automobile Schools," editorial, *Horseless Age* 18, 2 May 1906, 625–26. See also R.D.B., "Fake Automobile Schools," *Horseless Age* 17, 28 March 1906, 484; "Teaching the Chauffeur," *Automobile Dealer and Repairer* 1, July 1906, 184; P. S. Tice, "Automobile Schools," *Horseless Age* 19, 8 May 1907, 623–24; P. S. Tice, "Automobile Schools—Their Purposes and Methods," *Horseless Age* 21, 18 March 1908; Arthur Louis Glor, "Good Chauffeurs Needed," *New York Times,* 10 October 1910; and "N.A.D.A. Attacks Fake 'Schools,'" *Motor Age* 36, 11 December 1919, 15.

43. James D. Watkinson, "'Education for Success': The International Correspondence Schools of Scranton, Pennsylvania," *Pennsylvania Magazine of History and Biography* 120, October 1996, 343–69; and Watkinson, "Educating the Million: Education, Institutions, and the Working Class, 1787–1920" (Ph.D. diss., University of Virginia, 1995), 168–213.

44. Portland (Maine) Auto School brochure, 1907, Automobile Industry, box 20, folder 18, Warshaw Collection, NMAH.

45. *National Auto School,* brochure, 1913, Vertical File—Schools, Automotive History Collection, Detroit Public Library.

46. *Practical Auto School,* brochure, 1912.

47. *$1,000,000 Sweeney Automobile and Tractor School,* brochure, 1917.

48. Steve McIntyre quotes an International Association of Machinists (IAM) organizer's complaint that unionization of auto mechanics was difficult because "every second one of them has dreams of running his own garage." The IAM had organized only 2 percent of the nation's mechanics by 1928 and only 4 percent by 1935. Scattered other chauffeurs' and automobile mechanics' unions seem to have fared no better in organizing the trade. McIntyre, "'Repairman Will Gyp You,'" 466–67.

49. This certainly warrants investigation if similar records of other auto schools ever become available.

50. See enrollment figures reported in Education Committee Minutes, West Side YMCA, 10 March 1916.

51. "The University of Uncle Sam: All Colleges to Be Converted into Military Training Posts," *Engineering News-Record* 81, 29 August 1918, 420.

52. On America's relative unpreparedness for motorized warfare, see John C. Speedy III, "From Mules to Motors: Development of Maintenance Doctrine for Motor Vehicles by the U.S. Army, 1896–1918" (Ph.D. diss., Duke University, 1977), 1–174; James J. Flink, *The Automobile Age* (Cambridge, Mass.: MIT Press, 1992), 73–78; Erna Risch, *Quartermaster Support of the Army: A History of the Corps, 1775–1939* (1962; rpt., Washington, D.C.: U.S. Army, Center of Military History, 1989), 595–97; and W. F. Bradley, "Organization of the French Army Automobile Service," *Automotive Industries* 39, 26 December 1918, 1093–95.

53. Speedy, "From Mules to Motors," 136.

54. See Francis H. Pope, *Notes on the Organization and Operation of a Motor Truck Company* (Fort Sam Houston, Tex.: n.p., 1917); and "How the U.S. Army Truck Co. Is Organized and Operated," *Commercial Vehicle* 17, 1 August 1917, 24–25.

55. U.S. War Department, *War Department Annual Reports, 1917* (Washington, D.C.: GPO, 1918), 313–15; Victor W. Page, "Substituting Gasoline for Horseflesh: Work of Motor Trucks with the Army in Mexico," *Scientific American* 115, 5 August 1916, 118–19; U.S. Coast Artillery School, *Military Motor Transportation* (Fort Monroe, Va.: Coast Artillery Journal, 1923), 2; Speedy, "From Mules to Motors," 175–204; and Risch, *Quartermaster Support,* 597–98.

56. U.S. War Department, *War Department Annual Reports, 1917;* Page, "Substituting Gasoline for Horseflesh"; U.S. Coast Artillery School, *Military Motor Transportation;* Speedy, "From Mules to Motors"; and Risch, *Quartermaster Support.* Also see "Chauffeurs Needed for Army Now," *New York Times,* 19 November 1917; and Ludlow Clayden, "Must Train Military Truck Drivers . . . More Trucks than Drivers," *Automobile* 36, 19 April 1917, 768.

57. Reconstruction Parks, the largest of the overseas motor facilities established far behind the line of fighting, used skilled mechanics to perform the major repairing and rebuilding of army vehicles as well as manufacturing and salvaging of parts. Mechanics in Overhaul Parks also performed extensive repair work and occasionally overhauled vehicles, while those in the small, mobile Service Parks performed simple repairs and maintenance near the front line. Soldier-mechanics at Reception Parks, located at key ports, received and assembled crated vehicles sent from the United States and forwarded them

to the troops. For a discussion of how the army organized motor maintenance and repair operations at the various types of motor parks overseas, see Speedy, "From Mules to Motors," 365–66.

58. *A History of the Reconstruction Park 772, Motor Transportation Corps* (n.p.), 9–15, Unit History no. 1309–772, 1919, Military History Institute, Carlisle, Pa.; U.S. Adjutant General's Office, *The Personnel System of the United States Army* (Washington, D.C.: GPO, 1919), 1:27–31. Out of 425,000 men drafted through 15 December 1917, only 3,066 were considered journeymen or apprentice auto repairers, while 14,828 were comparably skilled with the care and handling of horses. See "Occupational Yield of Draft," Record Group 407, Records of the Adjutant General's Office, Committee on the Classification of Personnel, 1917–19, entry 283, box 13, National Archives, Washington, D.C.

59. *Personnel System of the United States Army,* vol. 1: 53–62.

60. See, e.g., U.S. Adjutant General's Office, Classification Division, *Personnel Specifications: Motor Transportation Corps* (Washington, D.C.: GPO, 1918).

61. Examples of the CCP's Trade Tests, including the performance test for truck drivers, are provided in U.S. Adjutant General's Office, *The Personnel System of the United States Army* (Washington, D.C.: GPO, 1919), 2:123–64.

62. On the military's use of psychological testing, see *Personnel System of the United States Army,* 2:219–31; Robert M. Yerkes, ed., *Psychological Examining in the United States Army,* Memoirs of the National Academy of Sciences, vol. 15 (Washington, D.C.: GPO, 1921); Clarence S. Yoakum and Robert M. Yerkes, *Army Mental Tests* (New York: Henry Holt, 1920; and Thomas G. Sticht and William B. Armstrong, *Adult Literacy in the United States: A Compendium of Quantitative Data and Interpretive Results* (San Diego, Calif.: San Diego Community College District, 1994), 27–49.

63. See Truman Lee Kelley, "Army Schools: Report of Interviews, with Recommendations, 8 January 1918," Record Group 407, Records of the Adjutant General's Office, Committee on the Classification of Personnel, 1917–19, entry 283, box 7, National Archives, Washington, D.C.

64. U.S. War Department, Motor Transportation Corps, *Report of the Chief of the Motor Transportation Corps to the Secretary of War* (Washington, D.C.: GPO, 1919), 7; "Brief History of the Holabird Quartermaster Depot and Its Activities" a fourteen-page typescript prepared after 1932, Record Group 92, Office of the Quartermaster General, Holabird Quartermaster Depot, 1919–33, 300.4 General Orders—354.2 Sales, box 2, folder 314.7, Military Histories, National Archives, Mid-Atlantic Region, Philadelphia. See also "Camp Holabird—Largest Truck Overhaul Depot," *Automobile and Automotive Industries* 39, 19 December 1918, 1052–55.

65. The West Side YMCA auto school began training army "chauffeurs" in January 1916 with a class of forty enlisted men but appears not to have been chosen by the CEST for more war training due to its lack of large dormitory facilities. Education Committee Minutes, West Side YMCA, 7 January, 10 March, 7 April, and 5 May 1916, YMCA/NY; and Frank E. Mathewson, "Final Report: Educational Director, Vocational Instruction, District 2 (New York, New Jersey) April to November 1918"; Record Group 165, Records of the War Department General and Special Staffs, Committee on Education and Special Training, entry 407, box 103, National Archives, Washington, D.C.; *Personnel System of the United States*

Army, 1:528–39; "Courses to Train Technicians and Mechanics for Army Service Begun," *Engineering News-Record* 80, 2 May 1918, 882–83; "Centralization of Education Functions of War Department," *Engineering News-Record* 81, 18 July 1918, 111; William T. Bawden, "Training the Fighting Mechanic," *Manual Training Magazine* 20, September 1918, 1–10; Penn Borden, *Civilian Indoctrination of the Military: World War I and Future Implications for the Military-Industrial Complex* (New York: Greenwood Press, 1989), 48–53.

66. During the late 1970s and early 1980s the Military History Institute (MHI) at Carlisle Barracks, Pa., conducted a survey of surviving World War I veterans. Vets were asked to fill out a multipage questionnaire and return it to MHI. Some respondents included keepsakes, diaries, photos, and other memorabilia from the war. David McNeal's diary can be found in his file in the World War I Survey Collection at the Military History Institute; hereafter cited as WWI Survey Coll., MHI.

67. U.S. Army, *C.C.P. Trade Test, Chauffeur, Truck Driver, 23-t, Performance* (Washington, D.C.: Committee on the Classification of Personnel, Trade Test Division, September 1918).

68. McNeal reported one fatality during the convoy, near Toledo, but did not elaborate on the cause. See McNeal diary entries for 11 July–26 August 1918, WWI Survey Coll., MHI.

69. This certificate, and what it represented, must have held great significance for David McNeal, as he kept it for sixty years before donating it, along with his diary, to the Military History Institute. McNeal's certificate is in his file, WWI Survey Coll., MHI.

70. A record of the range and extent of repair work performed in army service parks overseas can be found in the "Daily Reports of Jobs in Shop," Record Group 120, Records of the American Expeditionary Forces, WWI, Motor Transportation Corps, entry 1871, box 443, National Archives, Washington, D.C.

71. "Engineering Colleges Teach Fighting Mechanics for the Army . . . Automobile Mechanics Made in Eight Weeks," *Engineering News-Record* 81, 10 October 1918, 674–77. For profiles of other schools, see "War School for 100,00 Soldiers," *New York Times Magazine*, 28 April 1918; and "The University of Uncle Sam," *Engineering News-Record* 81, 29 August 1918, 420. See also "Outline of Vocational War Course at Old South Division High School, Chicago, Ill.," and "Outline of Vocational War Course at Purdue University, West Lafayette, Indiana," Record Group 165, Records of the War Department General and Special Staffs, Committee on Education and Special Training, entry 407, box 104, National Archives, Washington, D.C.

72. Stanley A. Zweibel, "Final Report: District Educational Director, Vocational Instruction District 3," 10, Record Group 165, Records of the War Department General and Special Staffs, Committee on Education and Special Training, entry 407, box 103, National Archives, Washington, D.C.

73. See Zweibel, "Final Report"; as well as the CEST final reports for districts 4 and 10, Record Group 165, Records of the War Department General and Special Staffs, Committee on Education and Special Training, entry 407, boxes 103–4, National Archives, Washington, D.C.

74. See, e.g., "Call for Motor Mechanics," *New York Times*, 18 January 1918; "Pershing Asks for Motor Mechanics," *New York Times*, 20 January 1918; "More 'Gas Hounds' Needed," *New York Times*, 14 October 1918; and "Call for 30,000 Workers: Quick Overseas Service for Men Handy at Motor Trades," *New York Times*, 30 October 1918.

75. U.S. Army, *The Personnel System of the United States Army*, 1:28; U.S. Army, Center

of Military History, *Order of Battle of the United States Land Forces in the World War, Zone of the Interior: Organization and Activities of the War Department* (1949; rpt., Washington, D.C.: U.S. Army, Center of Military History, 1989), vol. 3, pt. 1, 322.

76. See file of Herschel C. Hunt, Pvt., Co. B, Motor Repair Unit 315, WWI Survey Coll., MHI.

77. See file of Virgil R. Hertzog, PFC, Motor Transportation Corps, 310th Motor Repair Unit, WWI Survey Coll., MHI. See also "Motor Transportation Corps: Repair Unit 310, Third Army, Coblenz, Germany, 1918–1919," Unit History, MHI. Other sources indicate that military discipline was not a high priority in the motor repair units. *Let's Go,* the camp newspaper for the Reconstruction Park at Verneuil, France, informed readers in the spring of 1919 that they would have to be "remodeled from a mechanic to a doughboy" if they wanted to be discharged without delays when they arrived home. No more "slovenly, rheumatic salute." "The Colonel" (Col. Hegeman) offered a competitive drill to aid "in your evolution from mechanic to doughboy." Ironically, the winning section was treated to "a keg party at the chateau." See "Go Home One Month Earlier," *Let's Go,* 8 March 1919, 1–2; and "With One Foot on the Brass Rail," *Let's Go,* 3 May 1919, 6. These and three other issues of *Let's Go* can be found in the file of William H. Rumbaugh, Pvt., Motor Transportation Corps, Motor Repair Unit 301, WWI Survey Coll., MHI.

78. Letters found in file of Louis Chouinard, Cpl., Co. F, 439th Motor Supply Train, WWI Survey Coll., MHI. Dean Charles H. Snow of New York University, commenting on the automobile course offered there, declared that "many candidates for auto-mechanics and chauffeuring courses came to us with but little real knowledge of automobiles. Many could operate but knew nothing about the mechanism of their cars. The efficiency of all such men has been greatly increased. They [now] know about the mechanism, how to detect trouble, and how to repair. Many have a considerably increased earning capacity." Qtd. by Frank E. Mathewson in "Final Report, District Educational Director, Vocational Instruction, District 2," 5, Record Group 165, Records of the War Department General and Special Staffs, Committee on Education and Special Training, entry 407, box 103 National Archives, Washington, D.C.

79. Center of Military History, *Order of Battle,* 152, 269–70, 275, 342; "Y.M.C.A. Trucks Carry Movies to Sammies," *Commercial Vehicle* 17, 15 October 1917, 22–23; Virginia Scharff, *Taking the Wheel: Women and the Coming of the Motor Age* (New York: Free Press, 1991), 89–109; "Women's Motor Corps on Call Day and Night," *New York Times,* 7 April 1918.

80. See files for James K. Moore, PFC, Motor Transportation Corps, 312th Motor Repair Unit; Ernest A. Petrea, Pvt., Motor Transportation Corps, 312th Motor Repair Unit; and Oscar Arneberg, Sgt., Motor Transportation Corps, 309th Motor Repair Unit; WWI Survey Coll., MHI.

81. See file of Lynn F. Snoddy, Cpl., Motor Transportation Corps, 310th Motor Repair Unit, WWI Survey Coll., MHI.

82. See file of William H. Rumbaugh, Pvt., Motor Transportation Corps, Motor Repair Unit 301, WWI Survey Coll., MHI.

83. "Final Report, District Educational Director, Vocational Education, District 4," 18, Record Group 165, Records of the War Department General and Special Staffs, Committee on Education and Special Training, entry 407, box 104, National Archives, Washington, D.C.

84. "Final Report, District Educational Director, Vocational Education, District 5," 17, Record Group 165, Records of the War Department General and Special Staffs, Committee on Education and Special Training, entry 407, box 104, National Archives, Washington, D.C.

85. It might prove interesting for future research to follow these men and their activities in the postwar period, but that is not the immediate concern of this study.

86. "Army Motor Training for Expert Mechanics," *New York Times,* 10 August 1919.

87. "The Soldiers' Prospects," Motor Transportation Recruiting Circular no. 5, 1–2, Record Group 165, Records of the War Department General and Special Staffs, Committee on Education and Special Training, entry 310, box 419-A, National Archives, Washington, D.C.

88. "Tomorrow's Mechanics," *Motor Age* 38, 28 October 1920, 7–9.

89. The Federal Board, originally created by the Smith-Hughes Act of 1917 to foster civilian vocational education, was given charge of the vocational education of disabled soldiers in April 1918, with appropriations of two million dollars that year followed by fourteen million dollars the next. See Douglas, *American Apprenticeship,* 299–300. For a state-by-state breakdown of the 3,740 Federal Board students taking automotive-related courses in the fall of 1920, see "Getting Tomorrow's Mechanics . . . Men Being Trained by the Federal Board for Vocational Education," *Motor Age* 38, 16 September 1920, 12–14.

90. The Education Committee began noting enrollment decreases at its 11 November 1918 meeting, Armistice Day.

91. Education Committee Minutes, West Side YMCA, 9 December 1918, YMCA/NY.

92. Education Committee Minutes, West Side YMCA, 17 January 1919, YMCA/NY.

93. Education Committee Minutes, West Side YMCA, 7 February 1919, YMCA/NY.

94. Education Committee Minutes, West Side YMCA, 10 October and 6 November 1919, YMCA/NY.

95. See "Statement of Receipts and Expenses, 1920," Education Committee Minutes, West Side YMCA, 7 January 1921, YMCA/NY.

96. Mark J. Sweany, "Educational Work of the Knights of Columbus," in Department of the Interior, Bureau of Education *Bulletin,* no. 22 (1923): 4–5; L. Charles Anthony, "Seven Million Dollars: War Fund Being Used in After-War Welfare Work of Order in Reemployment, Education and Other Reconstructive Measures for Veterans," *Columbia,* April 1920, 7.

97. Six available photographs of mechanics and tire vulcanizing classes between 1919 and 1924 show no women students. See photos reprinted in the various reports of the Supreme Board of Directors, Knights of Columbus (K of C), Record Group SC-17-3, K of C Supreme Council Archives, New Haven, Conn.

98. The K of C maintained four schools in the South "exclusively for colored students"—in Birmingham, Jacksonville, Memphis, and Savannah—and reported that "such courses as tailoring, shoe repairing, embalming, band music, barbering, and elementary subjects are emphasized in response to the distinctive needs and abilities of this particular class of people." Auto mechanics was not mentioned, yet the National Automobile Chamber of Commerce included these four schools, in addition to sixty-four other K of C schools and thirty-two YMCA schools, in its list of automobile schools in 1922 and 1923. John J. Cummins, "Training Heads and Hands," *Columbia,* August 1921, 19; NACC, *Facts and Figures of the Automobile Industry,* for the years 1922 and 1923.

99. U.S. Bureau of the Census, *Thirteenth Census of the United States,* vol. 4: *1910,* U.S. Bureau of the Census, *Fourteenth Census of the United States Taken in the Year 1920,* vol. 4: *Population, 1920, Occupations.*

100. On 7 March 1920 the *San Francisco Chronicle* reported that the "insistent demand of several hundred former soldiers, sailors, and marines" prompted the K of C auto school in Oakland to open one month earlier than originally planned. On continuing high enrollments, see "Report of the Supreme Board of Directors, Knights of Columbus, War Work Activities, 1919," and "Report of the Supreme Board of Directors, Knights of Columbus, Educational and Welfare Work," for the years 1920, 1921, 1922, 1923, 1924, and 1925, Record Group SC-17-3, K of C Supreme Council Archives, New Haven, Conn.

Four • The Automobile in Public Education

1. The board postponed action on the request, and the earliest specific mention of an auto repair course was in 1922. The instructor of that course had been teaching machine shop since at least 1919, however, and may have included automobile work in the machine shop class. See George Ira Bohn, "History of Vocational Trade and Technical Education in San Bernardino to 1940" (master's thesis, Department of Industrial Education, California State University, Long Beach, 1973), 133; and the *Tyro,* the San Bernardino High School student annual for the years 1919–25, California Room, Norman Feldheym Library, San Bernardino, Calif.

2. "A Class in Vulcanizing Automobile Tires," *Industrial Arts Magazine* 5, February 1916, 89.

3. Similarly, Sacramento High School reported "capacity enrollment" in its new three-year auto mechanics course in 1921, and enrollment at the Automotive Trades School in Cincinnati grew from seventeen on opening night, 28 December 1918, to over one hundred by the end of the spring session. Lewis H. Wood, "Auto Mechanics in the High School[:] Development at Sioux City," *Industrial Arts Magazine* 13, November 1924, 421–23; "Auto-Mechanics at Sacramento, Calif.," *Industrial Arts Magazine* 10, May 1921, 201; and Ray F. Kuns, "Some Administrative Phases of an Automotive Trades School," *Industrial Arts Magazine* 12, March 1923, 85–89.

4. The Mechanics Club promoted "free discussion of many of the problems in connection with the work of the auto shop and [scheduled] addresses by prominent men of the automobile industry and other business men." See Riverside High School annual, *Orange and Green,* 1924, 144; and *Orange and Green,* 1925, 125, Local History Collection, Riverside Public Library, Riverside, Calif.

5. Charles Fleischman, "The Auto Mechanics Course in Consolidated Schools," *Industrial Arts Magazine* 12, June 1923, 239.

6. Harry W. Anderson, "A Course in Automobile Construction, Operation and Repair," *Industrial Arts Magazine* 9, September 1920, 343.

7. Education Committee Minutes, West Side YMCA, 13 February 1923, YMCA/NY.

8. Education Committee Minutes, West Side YMCA, 13 February, 14 May, 12 July, and 29 October 1923, YMCA/NY.

9. Education Committee Minutes, West Side YMCA, 9 May 1921, YMCA/NY.

10. A special report by the technical department of the United YMCA Schools suggested that the troubles at the West Side were due to an overemphasis on "owners and drivers courses" without better preparing "progressive, efficient and dependable mechanics." An examination of the West Side's finances by members of the International Committee criticized the branch for expecting that educational activities could be self-supporting and even income generating. That attitude, the report claimed, caused an "over-absorption of staff in finances and promotion [resulting in] marked limitations in personal guidance, placement, and character building." There was no doubt some merit to these criticisms, but the real root of the West Side's problems in the early 1920s lay elsewhere, in the addition of automobile curricula to public vocational school programs. See Education Committee Minutes, West Side YMCA, 19 June, 19 November, 28 December 1923, YMCA/NY; and "The Crisis in Automotive Schools," *Educational Messenger,* February 1923, 6–7.

11. The historical development of vocational education in the United States has been the subject of numerous studies. The unique factors influencing the development of automobile vocational education, however, have been overlooked. See Essay on Sources.

12. Excerpts of both John D. Runkle, "The Manual Element in Education" (1878), and Calvin M. Woodward, "The Fruits of Manual Training" (1883), can be found in *American Education and Vocationalism: A Documentary History, 1870–1970,* ed. Marvin Lazerson and W. Norton Grubb, Classics in Education, no. 48 (New York: Teachers College Press, 1974), 57–56. See also Violas, *Training of the Urban Working Class,* 125–27, 140–41; McClure et al, *Education for Work,* 20–26.

13. Katznelson and Weir, *Schooling for All,* 59.

14. Rorabaugh, *Craft Apprentice;* Violas, *Training of the Urban Working Class,* 17–66.

15. See, e.g., the often heated debate over manual training versus trade training which appeared in the pages of *American Machinist* from 1907 through 1920.

16. Lazerson and Grubb, *American Education,* 17.

17. The editor of *American Machinist* expressed the underlying antagonism many manufacturing interests held toward professional educators: "There is no body of men more incapable of handling trade schools successfully than existing boards of education." Keeping control of trade and vocational education out of the hands of educators was an important, though ultimately unrealized, goal for manufacturers. "Public Trade Schools," editorial, *American Machinist* 32, 30 December 1909, 1153. Excerpts of the National Association of Manufacturers, Committee on Industrial Education, reports from 1905 and 1912, can be found in Lazerson and Grubb, *American Education,* 88–100.

18. See American Federation of Labor, "Report of the Committee on Industrial Education" (1910), and "Report of the Commission on Industrial Relations" (1915), both excerpted in Lazerson and Grubb, *American Education,* 101–14; McClure et al., *Education for Work,* 51–52; and Katznelson and Weir, *Schooling for All,* 152.

19. See U.S. Federal Board for Vocational Education, *Bulletin No.1, Federal Board for Vocational Education: Statement of Policies* (Washington, D.C.: GPO, 1917).

20. In reality state and local school districts more than matched the federal funds they received because Smith-Hughes money covered only teacher salaries and teacher training. All of the additional investment in buildings and equipment for vocational education programs had to come from the state and local levels.

21. See U.S. Federal Board for Vocational Education, *Bulletin No. 4, Mechanical and Technical Training for Conscripted Men* (Washington, D.C.: GPO, 1918), 13, 18.

22. On Los Angeles schools, see "Final Report, District Educational Director, Vocational Instruction, District 11"; on Newark Central High School, see "Final Report, District Educational Director, Vocational Instruction, District 2," Record Group 165, Records of the War Department General and Special Staffs, Committee on Education and Special Training, entry 407, boxes 103 and 104, respectively, National Archives, Washington, D.C.

23. This school later became the South Division Continuation School offering auto mechanics courses to men employed in the trade. See "Final Report, District Educational Director, Vocational Instruction, District 7," Record Group 165, Records of the War Department General and Special Staffs, Committee on Education and Special Training, entry 407, box 104 National Archives, Washington, D.C.; and "Making Better Mechanics: What a Chicago School Is Doing to Help Train the Dealer's Mechanics," *Motor Age* 38, 18 November 1920, 14.

24. "Mechanics of Tomorrow: The Automotive Trades School of Cincinnati Is Giving Its Students a Course in Garage and Repair Shop Work," *Motor Age* 37, 20 May 1920, 10–11, 39; "Final Report, District Educational Director, Vocational Instruction, District 6"; "Final Report, District Educational Director, Vocational Instruction, District 7," Record Group 165, Records of the War Department General and Special Staffs, Committee on Education and Special Training, entry 407, box 104 National Archives, Washington, D.C.; and "Public Schools for Mechanics Recommended by Ohio Official," *Motor Age* 38, 18 November 1920, 27.

25. Flathead County Free High School in Kalispell, Mont., for example, is reported to have begun its Automobile Mechanics' department in 1917 and qualified for Smith-Hughes money but is not listed as training soldiers in the corresponding CEST district report. See "Tomorrow's Mechanics: How One High School in a Small Western Town Is Helping to Solve a Serious Problem for the Automotive Dealer," *Motor Age* 38, 15 July 1920, 15–16; and "Final Report, District Educational Director, Vocational Instruction, District 12," Record Group 165, Records of the War Department General and Special Staffs, Committee on Education and Special Training, entry 407, box 104, National Archives, Washington, D.C.

26. Fred Mason, "Ali Baba and His Forty Garage Thieves," *Illustrated World* 32, February 1920, 977–79, 1050.

27. "Twin Cities Garage Men Seek Better Mechanics," *Motor Age* 38, 1 July 1920, 35.

28. "Better Service Is Aim of Cooperative Education Campaign," *Motor Age* 39, 16 June 1921, 27.

29. "Apprentice Course Solves Labor Problem for Cleveland Distributor," *Motor World* 82, 19 March 1925, 24; and J. F. McDonald, "Training Repair Shop Personnel," *Journal of the Society of Automotive Engineers* 17 (August 1925): 202–4.

30. Numerous course outlines appeared in the vocational education literature in the 1920s, some of which are cited elsewhere in this chapter. Additionally, A. L. Dyke's *Automobile and Gasoline Engine Encyclopedia* (St. Louis: A. L. Dyke, 1910) remained widely available in regularly revised editions, and many private and public auto courses used it as a basic text. Dyke recorded sales of 8,176 of his books to private and public auto schools, agricultural colleges, and universities in 1926 alone. See Course Reports, No. Sold, etc., box 22, A. L. Dyke Collection Automotive History Collection, Detroit Public

Library. Hobbs and Elliott's book *The Gasoline Automobile,* first published by McGraw-Hill in 1915, was available in a revised second edition in 1919 and was also popular as an auto school textbook. During the war the Federal Board for Vocational Education published an "Emergency War Training Bulletin," with instructional outlines and necessary equipment for training gas engine mechanics. After the war two former instructors of the Motor Transportation Course at the Coast Artillery School, Edward S. Fraser and Ralph B. Jones, published *Motor Vehicles and Their Engines* (New York: D. Van Nostrand Co., 1919) as a textbook for the school market.

31. "Dealers—Get behind Your Public Schools," *Motor Age* 38, 22 July 1920, 18.

32. Wright also authored a widely used automotive textbook for high school auto shop use. "National Service Convention," *Journal of the Society of Automotive Engineers* 14, June 1924, 575–77. See also "Mr. Wright Becomes Federal Director," *Industrial Arts Magazine* 11, July 1922, 287; and advertisement for John Wiley and Sons, *Industrial Arts Magazine* 10, December 1921, xxvi.

33. "Recruiting Mechanics from Public High Schools," *Motor Age* 37, 15 April 1920, 10–11; and "Dealers and Technical Schools," *Motor World* 81, 2 October 1924, 33.

34. "Cleveland Wants Automobile Mechanics Taught in Schools," *Motor Age* 39, 3 February 1921, 37.

35. Katznelson and Weir, in *Schooling for All,* reveal the distinctly different local contexts of school reform in Chicago and San Francisco in the late nineteenth and early twentieth centuries, though they do not mention auto shop.

36. For descriptions of two industrial arts programs that included automobile work, see W. W. Patty, "Industrial Art Work at Berkeley, California," *Industrial Arts Magazine* 12, August 1923, 307–9; and R. G. Sawyer, "Industrial Arts in the Junior High School at Jacksonville, Fla." *Industrial Arts Magazine* 13, October 1924, 361–65.

37. Women certainly owned cars at the time, but the literature about automobile courses invariably spoke only of "fathers' " cars. This gendered language reflects an important part of the role that auto shop played in constructing the identity of auto mechanics, so I have not excised it.

38. Lewis H. Wood, "Auto Mechanics in the High School: Who and What to Teach," *Industrial Arts Magazine* 13, September 1924, 356–57.

39. J.W.S. Hodgdon, "Automobile Gasoline Engine Repair," *Manual Training Magazine* 20, June 1919, 344–45.

40. Charles Fleischman of Boise, Idaho, suggested that short evening courses "will bring the fathers into the school as nothing else can," thereby securing "very close community cooperation with the school." "The Auto Mechanics Course in Consolidated Schools," *Industrial Arts Magazine* 12, June 1923, 239.

41. Lewis, *Analysis of the Automechanic's Trade,* 30. Ray F. Kuns, "Some Administrative Phases of an Automotive Trades School—II: The Two Year Day Course," *Industrial Arts Magazine* 12, May 1923, 175–80.

42. Such job specialization never reached the degree seen on assembly lines of the period because the flow of repair work could not be easily predicted or regulated. Mechanics in departmentalized shops would likely have been shifted from department to department based on demand. See McIntrye, " 'Repair Man Will Gyp You,' " 266–76.

43. See Chevrolet Motor Co., *Chevrolet Service School Manual* (Detroit: Chevrolet Motor Co., 1925), 73–74; Auto Parts and Service Collection, Chevrolet (1912–27), HFMGV; and Ford Motor Co., *Service Mechanical Development* (Detroit: Ford Motor Co., 1935), 64–65, Accession No. 175, Ford Sales Literature, Service Mechanical Development (1935), HFMGV.

44. Pictured in the ad are "Body and Fender," "Paint," "Upholstery and Trimming," "Parts and Accessories," "Blacksmith," "Battery," "Electrical Laboratory," "Tire," "Chassis," "General Repair," and "Machine" departments. "Auto Trouble? Call Carl, Incorporated," advertisement clipping from the *Evening Star* (Washington, D.C.), 2 December 1927, Call Carl Files, Promotion Folder, Transp. Coll., NMAH.

45. McIntyre, "'Repair Man Will Gyp You,'" 130, 142–66. Instructions on how to fill out and use various forms comprised a major portion of the Packard Motor Co.'s instructions in *Packard Standard Service Methods for Packard Distributors and Dealers* (Detroit: Packard Motor Car Co., 1923).

46. From the early 1910s through the mid-1920s *Horseless Age, Motor Age, Motor World,* and other trade publications regularly featured articles on successful garages. Such articles very often included illustrations of the repair orders, time sheets, and other accounting forms used by those garages. Also, garage management books from the period, such as *How to Run a Retail Automobile Business at a Profit,* 2nd rev. ed. (Chicago: A. W. Shaw Co., 1924), gave examples of and instructions for using numerous accounting and tracking forms.

47. Examples of the forms used in a Milwaukee school are found in Paine, "Survey of the Boy's Technical High School of Milwaukee," 143–49.

48. See, e.g., Gerald A. Boate, "A Time and Cost Card System for the Mechanical Shops of the Newton Vocational School," *Industrial Arts Magazine* 5, August 1916, 332–34; N. F. Fultz, "The Making of an Alliance with a Chamber of Commerce," *Industrial Arts Magazine* 3, March 1915, 97–102; and Charles A. Carroll, "Hitching Vocational Education to Industry," *Industrial Arts Magazine* 13, August 1924, 292–95.

49. R. W. Hargrave, "Personal Efficiency or Forming Correct Habits," *Industrial Arts Magazine* 5, August 1916, 371.

50. The Wilbur Wright Cooperative School of Trades in Detroit specifically taught automotive students the "flat rate and bonus system" according to the school's assistant principal and Detroit's' assistant director of vocational education. Other explicit references to teaching flat rates in high school auto shops are lacking in the evidence gathered thus far. Earl L. Bedell and Frank Carpenter, "Cooperative Training in Automotive Maintenance," *Industrial Arts Magazine* 19, September 1930, 327–29.

51. Reminiscent of the Philadelphia Automobile Club's assessment of chauffeurs' character, Milwaukee auto shop students were also given overall ratings for their "Pupil Characteristics" of "Industry," "Attitude," and "Co-operation." Paine, "Survey of the Boy's Technical High School of Milwaukee," 143–49.

52. The California State Board of Education developed dozens of "Job Instruction Sheets" for the use of auto shop instructors and encouraged instructors to create more of their own. See Lewis, *Analysis of the Automechanic's Trade,* 38–115.

53. Los Angeles City School District, *Vocational Training Opportunities in the Los Angeles City High Schools,* School Publication no. 36, May 1921, National Archives, Record

Group 12, Office of Education, entry 82B, Federal Board for Vocational Education, Correspondence and Reports Concerning State Programs, 1917–37, box 6, California-Colorado, Folder California State Correspondence, July 1920–30 June 1921.

54. Lewis, *Analysis of the Automechanic's Trade,* 13, 34–36.

55. Paine, "Survey of the Boy's Technical High School of Milwaukee," 15–17, 85.

56. Precise evidence of average commercial repair shop size during the 1920s and 1930s is lacking, but Steve McIntyre cites several partial surveys indicating an average of anywhere from 3 to 8 mechanics per shop, while "one of Ford's largest service volume dealers in the 1930s . . . employed only forty mechanics in its service department." The independent Call Carl garage in Washington, D.C., claimed to employ "75 experienced mechanics" in 1927. Shop size thus ranged widely, but the more common experience was likely the smaller shop. McIntyre, "'Repair Man Will Gyp You,'" 138–40; "Auto Trouble? Call Carl, Incorporated." See also 1925 Call Carl photo showing 53 employees in front of shop and 1935 advertisement claiming "more than 100 employes [*sic*]," Call Carl Files, Promotion Folder Transp. Coll., NMAH.

57. On developments in vocational business skills courses, see McClure et al., *Education for Work;* and DeVault, *Sons and Daughters of Labor.*

58. John Younger, editor of *Automotive Abstracts,* wanted young mechanics to receive some business training but not to encourage independence. "Too many young fellows," said Younger, "see themselves get 60 cents an hour for their work and see the boss take in $1.20 and think that this is all velvet for the boss. A knowledge of overhead, equipment, sales expense and losses will change this opinion." Qtd. in J. Howard Pile, "Service and Engineering Convention Considers Timely Subjects," *Motor World* 81, 27 November 1924, 12–15.

59. Everett G. Glenn, "Establishing an Automotive Course," *Industrial Arts Magazine* 10, December 1921, 461–64.

60. Violas, *Training of the Urban Working Class,* 197.

61. Violas, *Training of the Urban Working Class,* 199. Parsons's book *Choosing a Vocation* was published by Houghton Mifflin in 1909.

62. "Vocational and Educational Guidance in Erie, Pa.," *Industrial Arts Magazine* 10, March 1921, 110–11. Three decades later Jacques Ellul criticized vocational guidance's tendency to discover in individuals "precisely the aptitudes which are essential to the needs of the capitalist economy. . . . It is the flimsiest make-believe to pretend that vocational guidance is in the service of human beings." *The Technological Society,* trans. John Wilkinson (1954; rpt., New York: Vintage Books, 1964), 359, 362.

63. Ray M. Simpson, "Selecting the Job for the Boy," *Industrial Arts Magazine* 13, March 1924, 90–93.

64. Meyer Bloomfield, "Vocational Guidance," *NEA Addresses and Proceedings,* 1912, 433, qtd. in Violas, 215.

65. Violas, *Training of the Urban Working Class,* 218.

66. George C. Greener, "Vocational Guidance in Prevocational Schools," *Industrial Arts Magazine* 9, October 1920, 381–83.

67. E. Joseph Goulart, "Some Uses of Intelligence Tests in Continuation School," *Industrial Arts Magazine* 11, October 1922, 400–401.

68. Edna Board et al., "The Relation of General Intelligence to Mechanical Ability," *Industrial Arts Magazine* 16, September 1927, 330–32; Payne Templeton, "Mental Testing and Vocational Counseling," *Industrial Arts Magazine* 19, April 1930, 132–35; Max S. Henig, "General Intelligence, Term of Stay and Trade Selected—Trade-School Students," *Industrial Arts Magazine* 19, October 1930, 367–69.

69. The attitudes expressed by Wood permeate educators' writings on auto shop during the 1920s, when these programs were being established and their purposes and objectives were being expressed. I have found no educators arguing that auto shop courses should be made more academically challenging or that they should be pitched at brighter students. Lewis H. Wood, "Auto Mechanics in the High School: Can We Justify It?" *Industrial Arts Magazine* 13, July 1924, 275.

70. Harry W. Anderson, "A Course in Automobile Construction, Operation and Repair," *Industrial Arts Magazine* 9, September 1920, 343–46.

71. See Paine, "Survey of the Boys' Technical High School of Milwaukee"; Lewis, *Analysis of the Automechanic's Trade;* Oakland Public Schools, *Automobile Repair in the Vocational Continuation School,* Superintendent's Bulletin no. 33, Course of Study Series (Oakland, Calif.: Oakland Public Schools, 1922); Lewis S. Neeb, "The Automobile as a Subject of Instruction in the Public Secondary Schools" (master's thesis, University of Arizona, 1927); and Lynn C. McKee, "A Trade Training Curriculum in Automobile Mechanics for Senior High Schools" (master's thesis, Duke University, 1931).

72. George C. Greener made this comment concerning students entering the vocational schools of Boston. "Vocational Guidance in Prevocational Schools," 382.

73. Kuns, "Some Administrative Phases of an Automotive Trades School—II," 180. A detailed examination of student records in several school districts would likely produce similar findings.

74. Scharff, *Taking the Wheel,* 108–9; "Order of the Winged Wheel," *Motor,* December 1921, 35.

75. The Smith-Hughes Act required states seeking matching funds to file annual statistical reports to the Federal Board for Vocational Education at the conclusion of each academic year. These reports detailed which schools offered which classes, how many students of each gender were enrolled in each class, how many teachers taught each subject, and how much money the state sought in relation to each course in their vocational, agricultural, and home economics programs. Beginning in 1920, states with racially segregated schools were required to report statistics separately for those schools; racial statistics for students can therefore be at least partially recovered. I sorted through all forty-eight states' reports for the sample years 1920 (in five cases in which the 1920 report was not available, I used the report of the next closest year), 1930, and 1937 (the last year available). I recorded each of the 44,598 students enrolled in each type of automobile mechanics course and created a database and spreadsheet from which I could derive the observations detailed herein. See Record Group 12, Office Education, entry 82, Records of the Assistant Commissioner for Vocational Education, Vocational Education Division, State Files, 1917–37, National Archives, College Park, Md.

76. Lewis H. Wood, "Auto Mechanics in the High School: Who and What to Teach," *Industrial Arts Magazine* 13, September 1924, 356–57.

77. Darlene Kirk's experience nearly a half-century later attests to the tenacity of gender barriers in shop classes generally. In 1971 she wanted to take drafting and shop classes at her high school in Houston. "But they never had any girls in the classes," Kirk recalled. "When I requested those classes, both my parents had to come down to the school and talk with the principal." Together they persuaded him that she needed the classes because she planned to open a craft shop someday. "So in 1971, I was the first female in the Houston school district allowed in the shop classes." Qtd. in Marilyn Root, *Women at the Wheel: 42 Stories of Freedom, Fanbelts, and the Lure of the Open Road* (Naperville, Ill.: Sourcebooks, 1999), 67.

78. Clyde W. Hall, *Black Vocational, Technical and Industrial Arts Education: Development and History* (Chicago: American Technical Society, 1973), 173–78. See also Ambrose Caliver, *Vocational Education and Guidance of Negroes: Report of a Survey Conducted at the Office of Education* (1937; rpt., Westport, Conn.: Negro Universities Press, 1970).

79. Kathleen Franz, " 'The Open Road': Automobility and Racial Uplift in the Interwar Years," in *Technology and the African-American Experience: Needs and Opportunities for Study*, ed. Bruce Sinclair (Cambridge, Mass.: MIT Press, 2004), 131–53.

80. The separateness of this black automobile economy even extended to racing, a window onto which Todd Gould has recaptured in *For Gold and Glory: Charlie Wiggins and the African-American Racing Car Circuit* (Bloomington: Indiana University Press, 2002). With the exception of Franz's study and Gould's, the extent of this black automobile economy has been largely overlooked by historians. One explanation for this oversight can, of course, be found in the invention and production orientation of traditional automotive historiography. Another explanation can be found in black scholars' earlier dismissal of black business entrepreneurs as merely unenlightened petit bourgeoisie who took advantage of black employees and customers. Finally, scholars who have looked at black businesses have been interested in white-collar, professional entrepreneurship rather than greasy garage and service station owners. See Robert E. Weems Jr., "Out of the Shadows: Business Enterprise and African American Historiography," *Business and Economic History* 26 (Fall 1997): 200–212.

81. J. H. Harmon Jr., "The Negro as a Local Business Man," *Journal of Negro History* 14 (April 1929): 116–55.

82. Harmon mentioned two "automobile agencies" in Kansas City and Chicago "owned and operated by Negroes" in 1929. It is unclear, however, whether they were new or used car dealerships. Harmon, "Negro as a Local Business Man."

83. The U.S. Census's number of 1,679 black-owned auto business, which would translate into an average of 34 to 35 such businesses per state, might be lower than the actual number. One 1923 black business directory listed 1 "auto cleaners," 7 auto repairers, 1 auto garage, and the "Pyramid Automobile Association" in Chicago alone. Nonetheless, the census numbers reflect the relative importance of auto-related businesses in the black economy. U.S. Department of Commerce, Bureau of the Census, *Negroes in the United States, 1920–32* (Washington, D.C. :GPO, 1934), chap. 17: "Retail Business"; James N. Simms, comp., *Simms' Blue Book and National Negro Business and Professional Directory* (Chicago: James N. Simms, 1923), 64.

84. This racial concentration of automobile businesses, and of small businesses in

general, was likely evident in ethnic neighborhoods as well. Polish, Italian, and German immigrants might have also run garages that drew customers from ethnically homogeneous communities within larger cities. The discrimination they faced when motoring outside their communities would not have been as palpable or as extensive, however, as that faced by African Americans.

85. National Negro Business League, "Report of the Survey of Negro Business," Schomburg Center Clipping File, 1925–74, Business: Chronology, 1921–31.

86. The term *muddled middle* is from Grace Elizabeth Hale, *Making Whiteness: The Culture of Segregation in the South, 1890–1940* (New York: Vintage, 1998), 8–9.

87. John Dollard, *Caste and Class in a Southern Town* (1937; rpt., New York: Doubleday, 1957), 4–5; cited in Hale, *Making Whiteness,* 183–84.

88. Nate Shaw, *All God's Dangers: The Life and Time of Nate Shaw,* comp. Theodore Rosengarten (New York: Knopf, 1974), 251–52.

89. African Americans seem not to have patronized the black automobile economy simply because they had no choice. A 1937 survey indicates that they trusted black mechanics more than they trusted black professionals such as insurance companies, banks, or physicians. Only 2.8 percent of 107 Fisk college freshmen agreed that "the only time a Negro mechanic can be trusted to do a first rate job is when a white man is standing over him," and only 5.6 percent agreed that "the quickest way to ruin a good car is to turn it over to a Negro mechanic." By comparison, 83.3 percent of those same students agreed that "Negroes with more sense than race pride will take out insurance with white companies," 55.6 percent agreed that "any person honest with himself knows that Negro banks are not as safe as white banks," and 41.7 percent agreed that "if a member of my family were dangerously ill and I had the choice of either a white doctor or a Negro, no one could blame me for calling the white doctor." See Thomas E. Davies, "Some Racial Attitudes of Negro College and Grade School Students," *Journal of Negro Education* 6 (April 1937): 157–65.

90. See "Annual Statistical Report of the State Board for Vocational Education, State of Mississippi, to the United States Office of Education, Year Ended June 30, 1937" and "Annual Statistical Report of the State Board for Vocational Education, State of Minnesota, to the United States Office of Education, Year Ended June 30, 1937," RG12, entry 82a, boxes 117 and 120, National Archives, College Park, Md.

91. John C. Wright, assistant commissioner for Vocational Education for the U.S. Office of Education, told a Washington, D.C., radio audience in 1937: "The cooperative nature of the program under the Smith-Hughes Act, places certain limitations on the extent to which we can render assistance. . . . The Office of Education is interested in and desirous of cooperating with States in working out plans looking toward more adequate programs of vocational education for those groups not hitherto adequately served. Negroes constitute one of these special groups." It is significant, however, that beginning in the 1919–20 school year states maintaining racially segregated schools were required to report their use of federal funds for black and white schools separately, thereby creating a public record of the disparate affect of Smith-Hughes money. John C. Wright, "Social and Economic Changes as the Affect the Vocational Education of Negroes," typescript of radio address over NBC station WMAL, Washington, D.C., 10 November, 1937, Papers and Addresses of John C. Wright, vol. 10, 1935–37, Special Collections, Strozier Library, Florida State University, Tallahassee.

92. Hall, *Black Vocational, Technical, and Industrial Arts Education,* 181–82.

93. See "Annual Statistical Report of the State Board for Vocational Education, State of Arkansas, to the United States Office of Education, Year Ended June 30, 1930," RG12, entry 82a, box 017, National Archives, College Park, Md.

94. For a revealing insight into educators' view of black students' employment prospects, see Virginia Peeler, *The Colored Garage Worker in New Orleans* (New Orleans: High School Scholarship Association, 1929).

95. In addition to creating markets for the special curricula, textbooks, and education studies cited elsewhere in this chapter, auto shop programs became lucrative markets for tool and equipment manufacturers and for college programs to train auto shop teachers. See ads for the South Bend Lathe Works, *Industrial Arts Magazine* 9, June 1920, xxxi; Kansas State Teachers College, the Stout Institute, the Carnegie Institute of Technology, *Industrial Arts Magazine* 13, April 1924, xxxii–xxxiii; and the Bradley Polytechnic Institute, *Industrial Arts Magazine* 13, June 1924, xxxvii–xxxix.

96. Throughout the 1920s and 1930s most manufacturers and dealers continued to conduct training sessions for their working mechanics. Such classes ranged from one- or two-day sessions on performing certain repair operations within flat rate times, to twelve-week factory schools granting certificates of competence to graduates. Service bulletins, service letters, and repair manuals described new methods and equipment for mechanics both inside and outside the dealer networks. Those outside the factory/dealer networks could take advantage of training sessions and service schools offered by parts suppliers and component manufacturers. They could also take unit courses at public continuation schools, community colleges, and the remaining private auto schools. Exemplary of the merging of supplier and trainer roles for independent mechanics was the Ambu Engineering Institute in Chicago. Originally the makers of an automotive electric trouble-locating device, the American Bureau of Engineering opened a school for automotive electrical repair work in 1919. See "Ambu Starts Electrical School," *Motor Age* 36, 7 August 1919, 24; "Where Can I Get a Good Electrical Man?" *Motor Age* 38, 8 July 1920, 14; "Making Automotive Electricians: How One Chicago School Trains Men, Many of Whom Already Are Established in Business, to Do Better Work," *Motor Age* 40, 14 July 1921, 12–13; and school advertisement in *Automobile Dealer and Repairer* 39, March 1926, 90.

Five • Tinkering with Sociotechnical Hierarchies

1. On the differentiation between middle-class masculine values and working-class masculine values in the twentieth century, see Peter N. Stearns, *BE A MAN! Males in Modern Society* (New York: Holmes and Meier, 1979), esp. chaps. 4–5.

2. Don T. Hastings, "The Development of a Modern Service System," *Journal of the Society of Automotive Engineers* 12, February 1923, 193–204.

3. "Propose to License Competent Repairmen," *Motor World* 46, 16 February 1916, 39. See also "The Incompetent Repairman," editorial, *Motor Age* 28, 16 September 1915, 14; and "Law for Examining Mechanics," *Motor World* 57, 25 December 1918, 18.

4. "Buffalo Has Plan for Mechanics' License," *Motor Age* 36, 25 September 1919, 22.

5. On the licensing and regulation of other occupations at the state level, see Council

of State Governments, *Occupational Licensing Legislation in the States* (Chicago: Council of State Governments, 1952).

6. "Oregon to Classify Mechanics," *Motor Age* 36, 17 July 1919, 25; "Portland Mechanics Threaten Walkout: State License Examination Held to Blame for Organization of Men," *Motor Age* 36, 9 October 1919, 15

7. The certification program of Auto Mechanics' Lodge No. 289 is described in "Tacoma Mechanics Are Unionized," *Motor World* 56, 11 September 1918, 28–29; and "Examining Mechanics," *Motor World* 58, 12 February 1919, 29–30.

8. "Mechanics Licenses," editorial, *Motor Age* 36, 9 October 1919, 16.

9. "Why License Mechanics?" editorial, *Motor World* 66, 23 March 1921, 20.

10. Wellington Gustin, "Possible Laws Imposing Limitation upon Present Free Entrance into Garage Business," *Motor Age* 38, 2 December 1920; "Mechanics' Licensing Bills Meet Defeat in Four States," *Motor Age* 39, 24 March 1921, 25; "Service Managers Disapprove Move for Licensing Mechanics," *Motor Age* 39, 12 May 1921, 26; "Training Mechanics," *Journal of the Society of Automotive Engineers* 15, December 1924, 480.

11. Certification of automobile mechanics did not become a significant issue again until the 1970s, when the consumer rights movement and the environmental movement forged a powerful new coalition favoring certification. See chap. 7 of this study. On the debate over licensing during the Great Depression, see the series of articles by Stanley P. McMinn, editor of *Automotive Merchandising*, "Do Mechanics Want to Be Licensed?" July 1934, 16–17, 70–77; "Mechanics DO Want to Be Licensed" (August 1934): 16–17, 64–66; "Would Licensing Protect the Public?" (September 1934): 14–15, 90–94; "Would Licensing Better Conditions?" (October 1934): 30–31, 66–68. See also Joe Mansfield, "Yes, Licensing Would Help Us," *Automotive Merchandising*, November 1934, 62–64, 110–12; and Bob Garrett, "Horse-Sense vs. License," *Automotive Merchandising*, July 1935, 18–19, 74.

12. For contemporary examples of this engineering orientation toward production and reliability, see General Motors Corp., Detroit, *The Scientific Side of the Automobile*, brochure presented to attendees at the SAE convention, Detroit, 27 June 1912; and C. F. Kettering, *Science and the Future Automobile: Being a Talk Given by C. F. Kettering to the Society of Automobile Engineers, at Their Annual Summer Meeting on Board the* S. S. Noronic, *June 18, 1916*. Both brochures in Vertical Files—Automobiles—Research and Testing, Transp. Coll., NMAH. On new steel alloys in the auto industry, see Thomas J. Misa, *A Nation of Steel: The Making of Modern America, 1865–1925* (Baltimore: Johns Hopkins University Press, 1995), chap. 6.

13. T. F. Cullen, "Need for Greater Service Accessibility in Car Design," *Society of Automotive Engineers Journal* 8, March 1921, 257–64. B. M. Ikert, "A Service-Man's Critical Estimate of Automotive Engineering," *Society of Automotive Engineers Journal* 10, April 1922, 263–69; "Discussion of Papers at the Chicago Service Meeting: A Service-Man's Critical Estimate of Automotive Engineering," *Society of Automotive Engineers Journal* 11, August 1922, 173–79.

14. P. J. Durham, "Problems of Motor-Vehicle Electrical-Equipment Maintenance," *Society of Automotive Engineers Journal* 14, February 1924, 162–64. See also A. H. Packer, "Maintenance Effects of Automotive Electrical-Equipment Standardization," *Society of Automotive Engineers Journal* 12, March 1923, 283–85; F. E. Moskovics, "Two Kinds of En-

gineering," *Society of Automotive Engineers Journal* 13, July 1923, 37–38; J. W. Lord, "The Factory Engineering Staff and Its Relation to Service," *Society of Automotive Engineers Journal* 13, July 1923, 93–100; J. W. Tracey, "The Engineer's Duty in Simplifying Electrical-Equipment Repairs," *Society of Automotive Engineers Journal* 14, January 1924, 11–12; B. C. Hinkley, "Design Standardization for Simplified Service," *Society of Automotive Engineers Journal* 18, March 1926, 268–70.

15. Cook quoted in discussion following J. W. Lord, "The Factory Engineering Staff and Its Relation to Service," 100. O. T. Kreusser, "Providing the Engineer with Service Data," *Society of Automotive Engineers Journal* 14, January 1924, 63–73; A. E. Hunt, "Needed Relations between Service Station and Factory," *Society of Automotive Engineers Journal* 19, August 1926, 167–72, 182; and *General Motors Corporation Research Laboratories, Detroit, Michigan,* brochure, Vertical Files, Automobiles, Research and Testing, Transportation Collection, NMAH.

16. "Chronicle and Comment," *Society of Automotive Engineers Journal* 14, May 1924, 467. Bachman and Gorey both quoted in "Discussion of Papers at Chicago Service Meeting," 174. If sufficient sources can be found, the service design movement of the 1920s should provide a useful case study of the social construction of automotive technology. I have not pursued this investigation further as it is tangential to my main argument about the creation and definition of mechanics. The Social Construction of Technology (SCOT) model is described in Trevor J. Pinch and Wiebe E. Bijker, "The Social Construction of Facts and Artifacts: Or How the Sociology of Science and the Sociology of Technology Might Benefit Each Other," in *The Social Construction of Technological Systems,* ed. Wiebe E. Bijker, Thomas P. Hughes, and Trevor Pinch (Cambridge, Mass.: MIT Press, 1987), 17–50; Wiebe E. Bijker, *Of Bicycles, Bakelites, and Bulbs;* and Trevor Pinch, "The Social Construction of Technology: A Review," in *Technological Change: Methods and Themes in the History of Technology,* ed. Robert Fox (Amsterdam: Harwood Academic Publishers, 1996).

17. "Flat-Rates Forcing Accessibility," *Society of Automotive Engineers Journal* 14, May 1924, 468. McIntyre makes this argument in his dissertation, "'Repair Man Will Gyp You,'" 178–82.

18. Clothing analogy attributed to a Mr. Cobleigh in Don T. Hastings, "The Development of a Modern Service System: The Discussion," *Journal of the Society of Automotive Engineers* 12 (June 1923): 539–44.

19. McIntyre, "'Repair Man Will Gyp You,'" esp. chaps. 3 and 5.

20. Ralph C. Rognon, "Service the Present Motor Car Problem," *Motor Age* 36, 25 December 1919, 19.

21. Hastings, "Development of a Modern Service System," 195; and "Development of a Modern Service System: The Discussion," 542.

22. Paul Dumas, "Where Flat Rate Stands Today," *Motor Age* 47, 22 January 1925, 34.

23. Hastings, "Development of a Modern Service System: The Discussion," 542.

24. Hastings, "Development of a Modern Service System: The Discussion," 540.

25. McIntyre, "'Repair Man Will Gyp You,'" 506–18; and McIntyre, "The Failure of Fordism: Reform of the Automobile Repair Industry, 1913–1940," *Technology and Culture* 41, April 2000, 269–99.

26. I can recall from my childhood in the early 1970s that mechanics in my father's tire

shop used times in published flat-rate volumes as starting points to be rounded up from depending on the condition of the car and the relationship with the customer. I have not found any direct evidence of flat-rate times being used in this manner during the 1920s, but it seems a reasonable extrapolation.

27. McIntyre, "'Repair Man Will Gyp You,'" 518.

28. Ford, *Service Mechanical Development*, 12.

29. By the 1950s all the major U.S. automakers still compiled flat-rate manuals and would only reimburse their dealerships' service departments for new car warranty work based on those times. See chap. 6 of this study.

30. "Make the Laboratory Test Set Your Ace Sales Builder," *Ford Service Merchandising Bulletin* 1, February 1935, 5–6.

31. For an illustration of the Ambu Electric Trouble Shooter, see the Co.'s full-page advertisement in *Automobile Dealer and Repairer* 29, April 1920, 2. Illustrations and descriptions of the Testall Trouble Finder, and the Niehoff Defectometer, appear in the Service Equipment section of *Motor Age* 36, 31 July 1919, 52 and 6 November 1919, 105, respectively. The Weston Fault Finder is featured in a Western Electrical Instrument Co. advertisement in *Automobile Dealer and Repairer* 34, October 1922, 12.

32. The Weidenhoff test bench is featured in "New Accessories at the Show," *Motor Age* 38, 25 November 1920, 30–31. Niehoff's test bench is illustrated in that company's advertisement in *Automobile Dealer and Repairer* 34, October 1922, 6.

33. See Service Equipment section of *Motor Age* 36, 25 December 1919, 49.

34. Chevrolet Motor Co., *Chevrolet Electrical Repair Manual for Four-Ninety and Superior Models* (1924; rpt., Detroit: Chevrolet Motor Co., 1978). Auto Parts and Service Collection, box: Chevrolet (1912–27), HFMGV.

35. "Convincing Owners," *Motor Age* 38, 30 September 1920, 14–15.

36. See "Sell Your Service to Your Customer the First Time He Comes In," *Motor Age* 38, 23 December 1920, 9; and "Eliminating Guesswork Detecting Electrical Troubles," *Motor Age* 38, 30 December 1920, 13.

37. Joseph Wiedenhoff, Inc., Chicago, *Catalog No. 45*, 1 January 1936. See statements made of their Model 830, Model 828, and Model 842 Motor Analyzers in Trade Catalog Collection, NMAH.

38. Sun Manufacturing Co., typescript manual for Sun Motor Tester, n.d., ca. 1933–35, in author's collection.

39. Stromberg Motoscope Corp., Chicago, *Automotive Testing Instruments*, 1933, in Trade Catalog Collection, NMAH.

40. "Praises Ford Laboratory Test Set," *Ford Service Merchandising Bulletin* 1, May 1935, 2. A 1935 letter to dealers told of improvements made to the Test Set since its introduction the previous year, which included doubling the size of the unit's meters. Service Department letter, "Improved Ford Laboratory Test Set," 27 March 1935, Accession no. 235, box 72, General Letters January–May 1935, HFMGV.

41. The "Bean Wheel Aligner" made by the K. R. Wilson Co. for Ford dealers used a very large indicator gauge to show customers when their front wheels were out of alignment. See "Bean Wheel Aligner Earns $11,700 in One Year," *Ford Service Merchandising Bulletin* 2, February 1939, 4; and promotional photo of unit, negative no. P833.64316-C. HFMGV.

42. On numbers of units sold in the United States, as well as overseas, see letter from Service Department, "All Foreign Branches," 11 October 1935, Accession no. 235, box 72, General Letters. On naming of Ford's diagnostic unit, see letter from Service Department, "Parts Sales," 19 December 1934, Accession 235, box 73, General Letters, HFMGV.

43. Compare advertisements, "When We Were Younger," Accession no. 19, box 115, folder Dealer Services, Service, 1931; and "Right Fit Is Important," Accession no. 19, box 117, folder Q-R, HFMGV.

44. Letter to dealers, "Parts and Service Promotional Activity," 17 January 1935, Accession no. 235, box 72, HFMGV.

45. Shoshana Zuboff describes this type of knowledge as sentient, embodied, experience-based knowledge. Michael Polanyi, *The Tacit Dimension* (London: Routledge and Kegan Paul, 1967); Douglas Harper, *Working Knowledge: Skill and Community in a Small Shop* (Chicago: University of Chicago Press, 1987); Shoshana Zuboff, *The Age of the Smart Machine: The Future of Work and Power* (New York: Basic Books, 1988).

46. "Big Field for Good Chauffeurs," *New York Times,* 19 November 1905.

47. A. J. Brennan, "What Sound Shows," *Automobile Dealer and Repairer* 15, July 1913, 43–44. See also "Locating Automobile Sounds," *Automobile Dealer and Repairer* 15, July 1913, 49; "Locating Pounds," *Motor Age* 36, 2 October 1919, 50–51; *Dyke's Automobile and Gasoline Engine Encyclopedia,* 6th ed. (St. Louis: A. L. Dyke, 1918), 638, 739; and description and illustration of the Knock Tector stethoscope manufactured by the E. R. Benson Mfg. Co., *Motor Age* 39, 2 June 1921.

48. For Kettering's characterization of rural garages, see "Kettering Analyzes Service Evils," *Journal of the Society of Automotive Engineers,* 13, December 1923, 523–24.

49. *Journal of the Society of Automotive Engineers* 17, December 1925, 518–19. John Squires and Carl Breer of the Maxwell-Chrysler Motor Corp. also describe "bothersome noises" in "Manufacturer's Reflections on the Automobile Service Field," *Journal of the Society of Automotive Engineers* 17, December 1935, 1 549–58.

50. Letters to branches, "Comparative Charts—Service Progress 1936," 22 April 1936; and "Comparative Charts Service Progress 1936," 22 July 1936, Accession 235, box 72, HFMGV.

51. Letter to branches, "1937 School Program," 19 January 1937, Accession 235, box 71, HFMGV; Ford Motor Co., *Service Mechanical Development,* 45–49.

52. Letter to branches, "Ford Laboratory Test Set Certificate," 5 May 1936, Accession 235, box 72, HFMGV; letter to branches, "1937 School Program."

53. Letter to branches, "Comparative Chart Service Progress 1936," 8 February 1937, Accession 235, box 71, HFMGV.

Six • Suburban Paradox

1. Anonymous Iowa mechanic interview quoted in Jim L. Drost, "Job Characteristics of Automotive Mechanics in Selected Iowa Dealerships and Garages" (Ph.D. diss., Iowa State University, Ames, 1970), 87.

2. NACC, *Facts and Figures of the Automobile Industry, 1941* (New York: NACC, 1941); and *Facts and Figures, 1948.*

3. Stephen E. Ambrose, *Citizen Soldiers: The U.S. Army from the Normandy Beaches to the Bulge to the Surrender of Germany, June 7, 1944–May 7, 1945* (New York: Simon and Schuster, 1997), 64–65. On the overall significance of U.S. and Allied industrial capacity to the outcome of World War II, see Alan S. Milward, *War, Economy, and Society, 1939–1945* (Berkeley: University of California Press, 1977).

4. For a brief description of the one-month and three-month courses offered at Holabird, see "Motorization Means More Mechanics . . . So the Army Enlarges Its Auto School," *Popular Science* 39, December 1941, 138–40.

5. The annual old-car scrappage rate declined during the war years to 40 percent of the prewar rate, while the average age of vehicles in use increased by 64 percent. Flink, *Automobile Age*, 275; Rae, *American Automobile*, 152; *Facts and Figures*, 1946–47.

6. "Repair Shop Manpower Situation Checked by American Automotive Association," *Motor Service*, 15 March 1943, 72. See also Martin Bunn, "Gus Keeps 'Em Rolling," *Popular Science*, December 1941, 142–44, 218. *Business Week* reported a greater loss of mechanics among dealerships in "Dealers Worried: Auto Men Have Lost 75% of Their Mechanics," 26 December 1942, 24+.

7. Studebaker Corp., *Women for Automotive Maintenance Service: A Manual for Dealers and Service Men* (South Bend, Ind.: Studebaker Corp., 1943), 5.

8. Evelyn Rand, interview by Rita Breton, Bangor, Maine, 5 April 1982, transcript, Northeast Archives of Folk and Oral History, Maine Folklife Center, University of Maine, Orono, 11, 15–16.

9. Rand, interview, 9, 12, 13, 24–25.

10. Ibid., 17 March 1982, 26, 32–33. Rand's determination to stay in the industry eventually landed her a job in a dealership parts department, where her duties included filing customers' service records. With the shop door once again closed to her, she became interested in bookkeeping and took night school and correspondence courses in bookkeeping and accounting. About six years after the war, she left the automotive service industry to become payroll clerk at the Eastern Maine Medical Center and advanced into another male-dominated field as supervisor of the hospital's new IBM Processing Center.

11. Employment figures derived from 1940, 1950, and 1960 U.S. Census data.

12. For more on women in the wartime manufacturing workforce and their experience afterward, begin with Ruth Milkman, *Gender at Work: The Dynamic of Job Segregation by Sex during World War II* (Urbana: University of Illinois Press, 1987), esp. chap. 7.

13. The Labor Department based its estimate in part on projected growth of motor vehicle registrations at prewar levels. The number of private and commercial cars, trucks, and buses registered in the United States increased at just over 20 percent from 1930 to 1940, so they reasoned that, with the end of depression and war, registrations would again increase by an additional 22 percent by 1950 to 39.1 million motor vehicles. They also reasoned that if the ratio of cars to mechanics remained about the same, at 85 registered vehicles per employed mechanic, the economy would support 450,000 to 460,000 mechanics by 1950. "Employment Outlook for Automobile Mechanics," *Monthly Labor Review*, February 1946, 211–27. Also available as U.S. Department of Labor, *Bulletin No. 842: Employment Outlook for Automobile Mechanics* (Washington, D.C.: GPO, 1945).

14. "Employment Outlook," 220.

15. Ibid., 221. See also Franz Serdahely, "Number of Mechanics Increasing," *Motor Age* 65, December 1945, 20–21, 46, 50.

16. "How to Hire Returned Soldiers," *Motor Age* 65, February 1946, 32, 84, 86; "Solve Your Labor Shortage with Vets," *Motor Age* 65, October 1946, 20, 54, 58, 60.

17. U.S. Department of Commerce, *Establishing and Operating an Automobile Repair Shop* (Washington, D.C.: GPO, 1946), iii.

18. Two disabled veterans, for example, were reported to have secured start-up capital for their Quincy, Mass., shop from the Smaller War Plants Corp. The records of the Smaller War Plants Corp. and its post-1946 successor agencies at the National Archives would provide a good starting point for further study. See "Army Know-How Wins," *Popular Science Monthly*, December 1945, 82–84, 210, 214; and Jonathan Bean, *Beyond the Broker State: Federal Policies toward Small Business, 1935–1961* (Chapel Hill: University of North Carolina Press, 1996).

19. See "Filled Up," *Fortune*, January 1946, 225.

20. Vehicle registration numbers from U.S. Department of Transportation, Federal Highway Administration, *Highway Statistics Summary to 1995*, table MV-200, http://www.fwwa.dot.gov/ohim/summary95/mv200.pdf.

21. Flink, *Automobile Age*, 278–79. Flink does not cite his source for calculating U.S. production at two-thirds of world output in 1955, but drawing on U.S. Department of Commerce figures, the Automobile Manufacturers Association (AMA) published comparative production numbers that support Flink's claim. See AMA, *Facts and Figures*, 1956, 14.

22. Joseph Interrante, "You Can't Go to Town in a Bathtub: Automobile Movement and the Reorganization of Rural American Space, 1900–1930," *Radical History Review* 21, Fall 1979, 151–68.

23. Workforce numbers calculated from U.S. Census figures for 1950 and 1970. Registrations calculated from U.S. Department of Transportation, Federal Highway Administration, *Highway Statistics Summary to 1995*, table MV-200: "State Motor Vehicle Registrations, by Years, 1900–1995," http://www.fhwa.dot.gov/ohim/summary95/mv200.pdf,

24. U.S. Department of Commerce, Bureau of the Census, *Home to Work Travel Study*; cited in Automobile Manufacturers Association, *Facts and Figures* (Detroit: AMA, 1970), 53. A Federal Highway Administration study of data collected in 1969 and 1970 found that 67 percent of all automobile trips were for the purposes of "making a living" or "family business" such as shopping and medical/dental visits. U.S. Federal Highway Administration, Department of Transportation, *National Personal Transportation Study: Purposes of Automobile Trips and Travel*, report no. 10 (n.p., May 1974).

25. Some historians have recently noted that on certain parts of automobiles owners and mechanics could do less than in the 1920s and 1930s. Kathleen Franz makes the argument that automaker design changes in the 1930s thwarted some user modifications popular in the 1920s. Joe Corn has noted how the carburetor in particular was a contested site where designers attempted to prevent motorists and mechanics from making hasty adjustments that would affect the operation of the engine. Ron Kline and Trevor Pinch have shown how the Ford Motor Co. responded to rural Model T users by discouraging them from using their cars as stationary power plants and field tractors. Franz, *Tinkering: Consumers Reinvent the Early Automobile* (Philadelphia: University of Pennsylvania Press,

2005); Corn, "'Tinkeritis': Normal Automotive Practice or Virulent Social Disease? Tinkering with Cars, 1900–1940," paper delivered at meeting of the Society for the History of Technology, Baltimore, 16 October 1998; Kline and Pinch, *Consumers in the Country* and "Users as Agents of Technological Change."

26. For more on postwar hot-rodding and racing in America, see H. F. Moorhouse, *Driving Ambitions: An Analysis of the American Hot Rod Enthusiasm* (Manchester: Manchester University Press, 1991); Robert C. Post, *High Performance: The Culture and Technology of Drag Racing, 1950–2000* (Baltimore: Johns Hopkins University Press, 1994); William Carroll, *Muroc, May 15, 1938* (San Marcos, Calif.: Auto Book Press, 1991); Tom Medley and LeRoi Smith, *Tex Smith's Hot Rod History: Tracing America's Most Popular Automotive Hobby* (Osceola, Wis.: Motor Books International, 1990–94); Jerry Bledsoe, *The World's Number One, Flat-Out, All-Time Great, Stock Car Racing Book* (New York: Doubleday, 1974). On the growth of the aftermarket performance parts industry, see David Lucsko, "Manufacturing Muscle: The Hot Rod Industry and the American Fascination with Speed, 1915–1984" (Ph. D. diss., MIT, 2005).

27. Unless otherwise noted, biographical information and Yunick quotations are from vol. 1 of Yunick's three-volume memoirs, *Best Damn Garage in Town: The World according to Smokey* (Daytona Beach, Fla.: Carbon Press, 2001).

28. Neshaminy, Doylestown, and the Bucks County region did not report any automobile courses under the Smith Hughes Act in 1920, 1930, or 1937. Perhaps this contributed to his disenchantment with high school.

29. Yunick, *Best Damn Garage in Town,* 15–16, 20–21, 23–25.

30. *Highway Statistics Summary to 1995,* table MV-200. Automobile mechanic employment numbers derived from 1940 and 1950 U.S. Census data.

31. Yunick, *Best Damn Garage in Town,* 127–31.

32. Ibid., 128.

33. Henry McLemore, "He's All Ears for Engines," *Popular Science,* January 1964, 94.

34. Ibid., 192.

35. Ibid.

36. See, e.g., McCahill's review of the 1960 Chevrolet Corvair in *Mechanix Illustrated,* November 1959, 63–66, 178, 182, 186.

37. Pete Daniel, *Lost Revolutions: The South in the 1950s* (Chapel Hill: University of North Carolina Press, 2000), 95–96.

38. For Yunick's recollections of writing for *Popular Science* and *Circle Track* magazines, see Yunick, *Best Damn Garage in Town,* 268–72, 281–91.

39. William Carroll, "How to Take Care of a '57 Chevy," *Popular Science,* July 1957, 149–55. See also Carroll, "How to Take Care of a '57 Plymouth," *Popular Science,* August 1957, 181–87; and "How to Take Care of an Edsel," *Popular Science,* September 1957, 242–50.

40. "Spring Car Care: Do It Now—Don't Wait till Summer," *Popular Science,* April 1963, 158–62.

41. "You're on Your Own in This Garage," *Popular Science,* April 1951, 119. See also "Fix-It-Yourself Garage," *Popular Mechanics,* November 1951, 148–49; "Where Every Man's His Own Mechanic: U-Fix-Ur-Car, Co.," *Saturday Evening Post,* 20 January 1951, 88; and "Self-Service Repairs: U-Fix-Ure-Car [*sic*] Co." *Newsweek,* 17 December 1951, 82.

42. In addition to Yunick's and McCahill's columns, *Popular Mechanics* published a long-running feature entitled "Saturday Mechanic" and a brief series called "Trouble Shooting the Tough Ones," which served similar purposes.

43. "Say, Smokey—," *Popular Science,* March 1964, 68–69.

44. "Say, Smokey—," *Popular Science,* March 1964, 68–69; April 1964, 72–73; May 1964, 20–24.

45. "Say, Smokey—," *Popular Science,* October 1968, 14, 18, 22.

46. Ben Shackleford, "Masculinity, the Auto Racing Fraternity, and the Technological Sublime," in *Boys and Their Toys? Masculinity, Technology, and Class in America,* ed. Roger Horowitz (New York: Routledge, 2001), 229–50.

47. See testimony of William N. Leonard, professor of economics, Hofstra University, before the Senate Committee on the Judiciary, Subcommittee on Antitrust and Monopoly, *Automotive Repair Industry Hearings,* 90th Cong., 2nd sess., 1969–71, pt. 1, 39.

48. Lizabeth Cohen, *A Consumers' Republic: The Politics of Consumption in Postwar America* (New York: Vintage Books, 2003), 294. See also Paul C. Wilson *Chrome Dreams: Automobile Styling since 1893* (Radnor, Pa.: Chilton Book Co., 1976); David Gartman, *Auto Opium: A Social History of American Automobile Design* (London: Routledge, 1994); Karal Ann Marling, *As Seen on TV: The Visual Culture of Everyday Life in the 1950s* (Cambridge, Mass.: Harvard University Press, 1994).

49. William Winpisinger, a former mechanic turned general vice president of the International Association of Machinists and Aerospace Workers, recalled those years as the "dirty thirties," when mechanics "were so desperate for work" that they would "underbid one another and the one who was the hungriest could get the job by agreeing to do it for less pay than anyone else." Winpisinger, testimony, *Automotive Repair Industry Hearings,* pt. 1, 344. See also Marion Jennings, "Growing Up with the Automobile," interview by Rose D. Workman, Charleston, S.C., 10 February 1939, in American Life Histories: Manuscripts from the Federal Writers' Project, 1936–40, WPA Federal Writers' Project Collection, Manuscript Division, Library of Congress, http://memory.loc.gov/ammem/wpaintro/wpahome.html.

50. Automobile Manufacturers Association, *Automobile Facts and Figures* (Detroit: AMA, 1965), 67. AMA reported its source as "County Business Patterns," First Quarter, 1962, U.S. Bureau of the Census.

51. In 1950 the U.S. Census reported 37,996 black males employed as auto mechanics and repairmen. This represented 5.9 percent of all male auto mechanics that year, which was a lower representation than black males in the total workforce (8.6%) but higher than black males employed as craftsmen, foremen, and kindred workers (3.6%).

52. See R. Irving Boone, ed., *Negro Business and Professional Men and Women: A Survey of Negro Progress in Varied Sections of North Carolina,* vol. 2: *1946* (Wilmington, N.C.: R. Irving Boone, 1946), 10, 23, 34. A sense of the vast number of black-owned automobile businesses in communities across the United States can be gained by surveying the advertisements in publications such as Irving's and by surveying the entries in the various black motoring guides published from 1936 through the mid-1960s, including *The Negro Motorist Green-book, Travelguide, The Baker Handbook,* and *GO Guide to Pleasant Motoring.*

53. See John M. Gregory, "Success Can Come," *Crisis* 52, December 1948, 362–63, 380–81.

54. J. C. Penney, e.g., opened its first auto center in Melbourne, Fla., in August 1963

and its one hundredth auto center in Cleveland, Ohio, in September 1966. See address of Wilbert Durbin, Automotive Merchandising and Service manager, J.C. Penney, in U.S. Senate, Committee on Commerce, *Motor Vehicle Diagnostic Analysis Technology, 1971–85*, technical conference proceedings for the use of the Committee, 92nd Cong., 2nd sess., 1971, Committee Print, 43.

55. AMA, *Automobile Facts and Figures*, 1962, 35, 42.

56. Winpisinger, testimony, *Auto Repair Industry Hearings*, pt 1, 353. The success of these mass merchandiser auto centers in suburban markets gained the attention of the Ford Motor Co. as it pondered if and how to bolster its service sales in similar fashion. See intracompany communication, P. F. Lorenz to A. R. Miller, 30 August 1963, FMC Archives, AR-71-3, Executive Files: E. R. Laux, V.P. Marketing, 1960–66, box 2.

57. Warren J. McEleney, dealer and treasurer of the National Automobile Dealers Association (NADA), testimony, *Auto Repair Industry Hearings*, pt. 1, 371.

58. Malcolm R Lovell, deputy assistant secretary for Manpower and Manpower Administrator, Department of Labor, testimony, 19 March 1970, *Auto Repair Industry Hearings*, pt. 4, 1746–47.

59. See testimony of Winpisinger, *Auto Repair Industry Hearings*, pt. 1, 342; as well as Winpisinger's testimony before the Federal Trade Commission (FTC), "In the Matter of Auto Warranty Hearings," included in *Auto Repair Industry Hearings*, pt. 5, 2185–2200. During World War II the Bureau of Labor Statistics found in its survey of 3,083 dealer and independent repair shops with nine or more employees that "only 13% had union agreements covering substantial proportions of the employees." U.S. Department of Labor, "Employment Outlook for Automobile Mechanics," *Monthly Labor Review*, February 1946, 226.

60. See NADA, "Methods of Wage Payments Used for Mechanics," *Auto Repair Hearings*, pt. 5, app., 2285.

61. See, e.g., Harry Wright, past president of the Independent Garage Owners of America, testimony, *Automotive Repair Industry Hearings*, pt. 1, 304–9.

62. FTC, "Report on Automobile Warranties," included in *Automotive Repair Industry Hearings*, pt. 5, 2289–2419; quotes from 2409–11.

63. Warren McEleney, treasurer of NADA, testimony, *Automotive Repair Industry Hearings*, pt. 1, 368.

64. Yet when it came to more difficult electrical and accessory repair jobs, "the mechanics could book only 7.5 and 6.6 hours respectively in an 8 hour day." *New Car Warranty and Service Study*, NADA, July 1968; cited in FTC, "Report on Automobile Warranties," *Automobile Industry Hearings*, pt. 5, 119.

65. Winpisinger testimony before the FTC, *Auto Repair Industry Hearings*, pt. 5, 2187–88.

66. Seeing the error of their ways, they all began cutting back on their warranty coverage in 1968 and successive years. FTC, "Report on Automobile Warranties," *Automotive Repair Industry Hearings*, pt. 5, 2305–11.

67. McEleney, testimony, *Automotive Repair Industry Hearings*, pt. 1, 366.

68. Education Operations, TechRep Division, Philco Corp., "A Study of Ford Motor Co. Service Training Programs: Summary of the Report," FMC Archives, AR-71-3, Executive Files: E. R. Laux, V.P. Marketing, 1960–66, box 12.

69. Gladys Roth Kremen, "MDTA: The Origins of the Manpower Development and Training Act of 1962," available at U.S. Department of Labor Web site, www.dol.gov.

70. National Child Labor Committee, National Committee on Employment of Youth, Youth Work Program Review Staff, *Getting Hired, Getting Trained: A Study of Industry Practices and Policies on Youth Employment* (New York: National Committee on Employment of Youth, 1964), 112. Malcolm Lovell, deputy assistant director for Manpower, testimony, *Auto Repair Industry Hearings*, 1970, pt. 4, 1747–48.

71. Winpisinger, testimony, *Automotive Repair Industry Hearings*, pt. 1, 352, 356–59. Ford Motor Co. also entered into an agreement with the Department of Labor to administer a Manpower training program in 1965. See intracompany communication, from Service Operations Department, "Training Dealership Service Technicians: Manpower Development and Training Act Program Summary," 26 March 1965, FMC Archives, AR-71-05780, Executive Files: Wright Tinsdale, 1957–69, 1973, box 17.

72. Yon, testimony, *Automotive Repair Industry Hearings*, pt. 1, 312.

73. Mel Turner address, U.S. Senate Committee on Commerce, *Motor Vehicle Diagnostic Analysis Technology, 1971–85*, 79.

74. African-American mechanics still clearly suffered from discrimination in the industry. In the mid-1960s Columbia University researcher Bernard Levenson reported on the earnings disparity between black and white graduates of segregated vocational high school auto shop programs in Baltimore. Auto shop graduates from the all black Carver Vocational-Technical High School earned approximately 45 to 55 percent of what graduates from the 98 percent white Morganthaler Vocational-Technical High School earned, and this held true from a half-year after graduation through four and a half years after graduation. Levenson found no significant difference in parental education levels or in the graduates' levels of nonemployment and assumed that since both programs were high school level, there were "no evident factors which explain the differential earnings of the magnitude found . . . other than discrimination by industry." Yet as we have seen through the records produced by the Smith-Hughes Act in the 1920s and 1930s, auto shop programs at black and white schools often differed significantly in quality and curriculum. Responsibility for the income disparity in Baltimore could not be placed solely on the shoulders of industry; it was built into the structure of the automotive marketplace. Bernard Levenson and Mary S. McDill, "Vocational Graduates in Auto Mechanics: A Follow-up Study of Negro and White Youth," *Phylon: The Atlanta University Review of Race and Culture* 27, Winter 1966, 347–57.

75. Malcolm Lovell, testimony, *Automotive Repair Industry Hearings*, pt. 4, 1747. A 1957 study of auto shop students at Nicholas Senn High School in Chicago—a school in an area with few industrial plants, an above-average income bracket, and a 70–75 percent college entrance level—found that only thirteen of the thirty-five students listed "career" as the reason they were taking advanced auto shop courses. Five listed "part-time job." Far more listed maintaining their own cars, "hobby," or purchasing a car as their reasons for taking advanced auto shop. A. C. Vasis Jr., "Objectives of an Auto Shop Program" *Industrial Arts and Vocational Education* 46, September 1957, 208–12.

76. John Kushnerick, editor and publisher of *Motor Age*, testimony, *Automotive Repair Industry Hearings*, pt. 1, 144–45.

77. C. D. Crill, Advisory Board, California Bureau of Automotive Repair, testimony before the Senate Committee on Business and Professions, California State Legislature, *Public Hearing on Certification of Automotive Mechanics,* Los Angeles, 26 November 1974, 70; hereafter cited as *California Certification Hearing,* 1974.

78. Myron H. Appel, testimony, *California Certification Hearing,* 1974, 84–86. A 1970 study of Iowa automobile dealerships and garages found that 34 percent of the 242 mechanics questioned had not completed high school. This would be unusually low and calls into question whether the respondents were reluctant to admit their true level of schooling to an academic researcher. Nationally, the U.S. Census estimated the dropout rate among auto mechanics at 57 percent in 1970. For all male "craftsmen and kindred workers," the rate was 50 percent, and for all employed males over sixteen years old it was 42 percent. On Iowa mechanics, see Drost, "Job Characteristics of Automotive Mechanics in Selected Iowa Dealerships and Garages," 47. National figures are drawn from *1970 Census of Population,* vol. 2: *Subject Reports, 7A, Occupation Characteristics,* table 5.

79. National Child Labor Committee, National Committee on Employment of Youth, Youth Work Program Review Staff, *Getting Hired, Getting Trained: A Study of Industry Practices and Policies on Youth Employment* (New York: National Committee on Employment of Youth, 1964), 104–5.

80. Mel Turner address before the Motor Vehicle Diagnostic Analysis Technology Conference, in U.S. Senate, Committee on Commerce, *Motor Vehicle Diagnostic Analysis Technology, 1971–85,* technical conference proceedings for the use of the Committee, 92nd Cong., 2nd sess., 1971, Committee Print, 81.

Seven • "Check Engine"

1. Jackson successfully sued the belligerent customer for thirty-five hundred dollars to cover his medical expenses and lost wages. *Thomas Johnson v. Theodore Jackson,* District of Columbia Court of Appeals, 178 A.2d 327; 1962 D.C. App.

2. Bill Matters, testimony, California State Legislature, Senate Committee on Business and Professions, *Certification of Automotive Mechanics: Public Hearing, November 26, 1974,* (n.p., 1974), 140. Both Jackson's assailant and Matters seemed to feel the need to assert a physical, "alpha male" sort of threat, perhaps to overcome a supplicant/weak position of not knowing their own mechanical problems. Unable to live up to the mechanical masculinity typified by Smokey Yunick, each resorted to a more overt pugilistic masculinity to gain satisfaction.

3. Lizabeth Cohen, *A Consumers' Republic: The Politics of Mass Consumption in Postwar America* (New York: Vintage, 2004); Hayagreeva Rao, "Caveat Emptor: The Construction of Nonprofit Consumer Watchdog Organizations," *American Journal of Sociology* 103, January 1998, 912–61; Meg Jacobs, "'How about Some Meat?' The Office of Price Administration, Consumption Politics, and State Building from the Bottom Up, 1941–1946," *Journal of American History* 84, December 1997, 910–41; David A. Aaker and George S. Day, eds., *Consumerism: Search for the Consumer Interest* (New York: Free Press, 1978); Erma Angevine, ed., *Consumer Activists: They Made a Difference: A History of Consumer Action Related by Leaders in the Consumer Movement* (Mount Vernon, N.Y.: National Consumers Committee for Research and Education, 1982).

4. *Donald C. MacPherson v. Buick Motor Co.*, 153 A.D.474:138 N.Y.S. 224: 1912 N.Y. App. Div.

5. *Donald C. MacPherson v. Buick Motor Co.*, 160 A.D. 55; 145 N.Y.S. 462:1914 N.Y. App. Div; *Donald C. MacPherson v. Buick Motor Co.*, 217 N.Y. 382:111 N.E. 1050:1916 N.Y.

6. Class titles are drawn from, "Annual Statistical Report of the State Board for Vocational Education to the United States Office of Education, State of Florida, Year Ended June 30, 1937," Record Group 12, Office of Education, entry 82A, Records of the Assistant Commissioner for Vocational Education, Vocational Education Division, State Files, 1917–37, National Archives. These state reports detailing the courses and enrollment numbers for home economics courses are a rich, largely untapped resource for studies in this area.

7. Rao, "Caveat Emptor," 934, citing U.S. Bureau of the Census, *Historical Statistics of the United States.*

8. Cohen, *Consumers' Republic*, 30.

9. Warne, "Consumer Action Programs of the Consumers Union of the United States," in Aker and Day, *Consumerism*, 152–60; Rao, "Caveat Emptor," 929–30.

10. Colston E. Warne, "Consumer Union's Contribution to the Consumer Movement," in Angevine, *Consumer Activists*, 95.

11. Rao, "Caveat Emptor," 941–42.

12. Estimated readership in 1961 was four million. Robert O. Herrmann, "The Consumer Movement in Historical Perspective," in Aker and Day, *Consumerism*, 32–33.

13. Warne, "Consumer Union's Contribution," 101.

14. Rao, "Caveat Emptor," 945.

15. Examples of best-selling exposés include Mary C. Philips, *Skin Deep* (1934); and Arthur Kallett and Frederick Schlink, *100,000,000 Guinea Pigs: Dangers in Everyday Foods, Drugs, and Cosmetics* (1933).

16. Interestingly, Roger William Riis was the son of Jacob A. Riis, one of America's first photojournalists, whose 1890 book, *How the Other Half Lives*, documented the squalid living conditions among New York's slums and helped launch Progressive Era housing reforms. The younger Riis's exposé of shady and incompetent repair practices appeared as a series of *Readers Digest* articles: "The Repair Man Will Gyp You if You Don't Watch Out," July 1941, 1–6; "The Radio Repairman Will Gyp You if You Don't Watch Out," August 1941, 6-10; "The Watch Repairman Will Gyp You if You Don't Watch Out," September 1941, 10–12; and "The Repairman Will Gyp You if You Don't Watch Out: A Symposium," October 1941, 144–48.

17. For a sample of the reaction in the trade press, see "Crook, Saint, or Sucker?" *Motor Service*, 15 August 1941, 19–21, 64, 66, 90; "Typical Motor Service Reader Called On by 'Reader's Digest' Representative," *Motor Service*, 15 October 1941, 38, 40, 42; "Member of Society of Automotive Engineers Comments on Reader's Digest Article," *Motor Service*, 15 December 1941, 34, 38, 92.

18. This claim appears in the front matter of Roger Riis and John Patric, *Repair Men May Gyp You*, 2nd ed. (Florence, Ore.: Frying Pan Creek, 1949).

19. *Public Papers of the Presidents of the United States: John F. Kennedy, 1962* (Washington, D.C.: GPO, 1963).

20. Study quoted in Cohen, *Consumers' Republic*, 363.

21. See, e.g., the legislative history of the National Traffic and Motor Vehicle Safety Act (1966); and *The Automobile Insurance Study* (1968).

22. Michael O'Brien, *Philip Hart: The Conscience of the Senate* (East Lansing: Michigan State University Press, 1995).

23. Senate Committee on the Judiciary, Subcommittee on Antitrust and Monopoly, *Automotive Repair Industry Hearings*, 90th Cong., 2nd sess., 1969–71, pt. 4, 1559.

24. Roush had written these words in a letter to the editor of the *National Observer* in December 1968. He then forwarded a tear sheet of the published letter to Senator Ervin asking that it be entered into the *Congressional Record*, or at least into the record of "the hearings on auto repairs." *Automotive Repair Industry Hearings*, pt. 1, 623.

25. *Automotive Repair Industry Hearings*, pt. 1, 55.

26. *Automotive Repair Industry Hearings*, pt. 1, 363–64, 372–73.

27. By the end of the hearings Hart and his staff had settled on a figure of 30 percent. The actual number no doubt varied widely, yet a decade later, in 1979, University of Alabama researchers under contract to the U.S. Department of Transportation conducted a study of auto repair shop capabilities. Taking sixty-two different cars with induced malfunctions to a representative range of size and type of repair shops in seven major cities, they concluded that "the probability was 50–50 that the motorist had repairs made which were not needed or did not have the needed repairs made." See Burton L. Jones, Joseph F. Peters, and Bernard J. Schroer, *Selective Survey of the Capability of Representative Automobile Repair Facilities to Diagnose and Repair Automobiles*, a report prepared for the Department of Transportation, Consumer Affairs Division, May 1979; quote from 24.

28. *Automotive Repair Industry Hearings*, pt. 1, 422.

29. Ibid., 61.

30. Ibid., 83.

31. California Legislature, Senate Committee on Business and Professions, Public Hearing, *Certification of Automotive Mechanics*, Los Angeles, 26 November 1974, 7.

32. *Automotive Repair Industry Hearings*, pt. 6, app., 3610.

33. Ibid., pt. 1, 156.

34. Ibid., pt. 2, 52, 322, 889.

35. Booz, Allen and Hamilton, Inc., *Scope and Impact of New Automotive Technology on the Inspection, Diagnosis and Repair Process*, report prepared for the U.S. Department of Transportation, National Highway Traffic Safety Administration, November 1980, 2–2.

36. Paul Sutter, *Driven Wild: How the Fight against Automobiles Launched the Modern Wilderness Movement* (Seattle: University of Washington Press, 2002).

37. Frank Uekotter, "The Strange Career of the Ringelman Smoke Chart," *Environmental Monitoring and Assessment* 106 (2005): 11–26.

38. James Krier and Edmund Ursin, *Pollution and Policy: A Case Essay on California and Federal Experience with Motor Vehicle Air Pollution, 1940–1975* (Berkeley: University of California Press, 1977), 75–76.

39. Scott Hamilton Dewey, *Don't Breathe the Air: Air Pollution and U.S. Environmental Politics, 1945–1970* (College Station: Texas A&M University Press, 2000), 47–48; Krier and Ursin, *Pollution and Policy*, 79–80; Jack Doyle, *Taken for a Ride: Detroit's Big Three and the Politics of Pollution* (New York: Four Walls Eight Windows, 2000), 18–19.

40. Hugh S. Gorman uses the concept of legislatively or socially changing environmental boundary conditions affecting the design of technological systems, in *Redefining Efficiency: Pollution Concerns, Regulatory Mechanisms, and Technological Change in the U.S. Petroleum Industry* (Akron, Ohio: University of Akron Press, 2001).

41. For discussion of "oil dilution problem," see Chevrolet Motor Co., *Chevrolet Repair Manual: 1934 Master and Standard Models* (Detroit: Chevrolet Motor Co., 1935), 146–47. For comparative illustrations of Cadillac's positive crankcase ventilation versus the downdraft tube systems of Oldsmobile, Buick, and Packard, see Frank D. Graham, *Audel's New Automobile Guide for Mechanics, Operators and Servicemen* (New York: Theodore Audel & Co., 1947), 199–201. See also Ralph Nader, *Unsafe at Any Speed: The Designed-in Dangers of the American Automobile* (New York: Grossman, 1965), 156–58; and Doyle, *Taken for a Ride*, 508.

42. "Inexpensive Anti-Smog Device to Be Offered on Car Next Year," *New York Times*, 6 December 1959, 67.

43. Air injection systems remained mostly on manual transmission cars for which the engine speed at shifting was not predictable, making precise tuning for emissions difficult. U.S. Department of Health Education and Welfare, National Air Pollution Control Administration, *Control Techniques for Carbon Monoxide, Nitrogen Oxide, and Hydrocarbon Emissions from Mobile Sources* (Washington, D.C.: GPO, 1970), chap. 5.

44. Krier and Ursin, *Pollution and Policy*, 150–51.

45. A. Wiman, "A Breath of Death," transcript of special report on KLAC radio, Metromedia, Inc., Los Angeles, 1967, 45; cited in Krier and Ursin, *Pollution and Policy*, 152.

46. "40% of '66 Cars Fail California's Antismog Test," *New York Times*, 19 January 1967, 12.

47. Krier and Ursin, *Pollution and Policy*, 164.

48. Gladwin Hill, "California to Ban '68 Auto Sales if Smog Equipment Fails Tests," *New York Times*, 9 March 1967, 79.

49. Automobile Club of Missouri, "A Report on Defects in Automobiles and the Quality of Repair Work, May 1969," *Automotive Repair Industry Hearings*, pt. 6 app., 3033–3105.

50. On the reactions of automobile performance enthusiasts and aftermarket parts suppliers to increasing emissions controls, see H. F. Moorhouse, *Driving Ambitions: An Analysis of the American Hot Rod Enthusiasm* (Manchester: Manchester University Press, 1991), chap. 6.

51. HEW/NAPCA, *Control Techniques*, 4–6–7.

52. "Say, Smokey—," *Popular Science*, April 1972, 20.

53. Joe Corn, "'Tinkeritis': Normal Automotive Practice or Virulent Social Disease? Tinkering with Cars, 1900–1940," paper delivered at meeting of the Society for the History of Technology, Baltimore, 16 October 1998. The ideal, or stoichiometric, gasoline-air ratio is 14.7:1. This allowed the most efficient burn producing the least amount of hydrocarbons and carbon dioxide.

54. U.S. Environmental Protection Agency (EPA), Office of Enforcement and General Counsel, "Mobile Source Enforcement Memorandum No. 1a," 25 June 1974. For indications of the EPA's continuing concern with "tampering," see EPA, Mobile Source Enforcement Division, Technical Support Branch, *1978 Motor Vehicle Tampering Survey* (Washington, D.C.: GPO, 1979); and later annual surveys of similar title.

55. *Automotive Repair Industry Hearings,* pt. 4, 1735.

56. Ibid.; HEW/NAPCA, *Control Techniques,* 3–10.

57. Roy F. Knudsen, ed., *Vehicle Emission Measurement—Panel Discussion: Transcription of the Vehicle Emission Measurement Panel Presented at the Instrument Society of America Conference, October 26–29, 1970* (Pittsburgh: Instrument Society of America, 1971), 18.

58. Knudsen, *Vehicle Emission Measurement,* 18–19.

59. Hart's faith in a technological remedy to a complex socioeconomic problem was neither novel nor uncommon. Like Ford's hopes for its Laboratory Test Set in the 1930s, an "objective" engineered solution to the vexing service problem seemed preferable to other alternatives. Even today many place their hopes in such politically painless escapes from vexing problems. See the range of case studies in Lisa Rosner, ed., *The Technological Fix: How People Use Technology to Create and Solve Problems* (New York: Routledge, 2004).

60. "Remarks of Senator Philip A. Hart (Democrat-Michigan) to the Society of Plastics Engineers, Rackham Building, Detroit, Mich., January 19, 1970," rpt. in *Automotive Repair Industry Hearings,* pt. 4, 1558–62. On reaction to Hart's policy shift, see various correspondence concerning mechanic training and licensing, *Automotive Repair Industry Hearings,* pt. 6, app., 3598-3612; Donald A. Randal and Arthur P. Glickman, *The Great American Auto Repair Robbery* (New York: Charterhouse, 1972), 88–92. On the proposal of voluntary certification, see "Statement of National Automobile Dealers Association, Submitted to Senator Hart, March 19, 1970," rpt. in *Automotive Repair Industry Hearings,* pt. 4, 1769–71. See also the numerous national newspaper and magazine clippings generated by the launch of the National Institute of Automotive Service Excellence's (NIASE) certification program in 1972. Detroit Public Library, Automotive History Collection, Vertical Files, Servicemen.

61. For comparative analysis of state laws as well as industry opposition to statutory remedies at the state level, see National Association of Attorneys General, Committee on the Office of the Attorney General, *Legislation Regulating Auto Repair* (n.p.: National Association of Attorneys General, 1976); and Ruth W. Woodling, *Auto Repair Regulation: An Analysis,* 3rd ed. (Georgia: Institute of Government, University of Georgia, 1978).

62. The Swedish government publication *Weak Points of Cars* as well as *Consumer Reports'* "Frequency of Repair" feature provided models for Hart and his key staff counsel, Donald Randall.

63. Hart describes his vision for diagnostic centers in "Remarks of Senator Philip Hart," *Automotive Repair Industry Hearings,* pt. 4, 1558–62; and in U.S. Senate, Committee on Commerce, *Motor Vehicle Diagnostic Analysis Technology, 1971–85,* technical conference proceedings for the use of the committee, 92nd Cong., 2nd sess., 1971, Committee Print, 6–10.

64. Because the Hart Hearings also considered the topic of crash repairs, Title I of the Motor Vehicle Information and Cost Saving Act gave the secretary of transportation authority to establish vehicle bumper standards and is often referred to as "the bumper bill." Hart's main consumer interests are embodied, however, in Titles II and III of the act, which sought to compile and disseminate consumer information and to promote diagnostic test centers, respectively. This discussion incorporates Title III provisions of both the original 1972 act and its 1976 amendment. See U.S. Statutes at Large 86 (1972): 947; and 90 (1976): 981. Senator Hart also discusses the act in its pending Senate bill form in U.S. Senate, Committee on Commerce, *Motor Vehicle Diagnostic Analysis Technology,* 8–9.

65. See chap. 5 of this study.

66. Riis and Patric, *Repair Men May Gyp You,* 2nd ed. (1949), 52.

67. Automotive oscilloscopes represented one of the key equipment developments just before the first diagnostic centers. Beginning in about 1958, advances in the production of cathode ray tubes for the burgeoning television industry enabled automotive equipment makers to offer reasonably priced oscilloscopes to the service industry. Scopes offered some diagnostic advantages to mechanics who could now analyze the visual trace of ignition voltage pulses on the oscilloscope screen. It made electrical properties visual rather than strictly numerical, which may have appealed to some mechanics. A 1963 Ford internal study of diagnostic repair centers claimed that "the average experienced mechanic can be trained to use [an oscilloscope] with about 10 hours of instruction." A 1978 Department of Transportation study noted, however, that an oscilloscope required "more extensive training . . . to interpret the graphical patterns" than did conventional meter-based equipment. See intracompany communication, P. F. Lorenz to A. R. Miller, 30 August 1963, FMC Archives, AR-71-3, Executive Files: E. R. Laux, V.P. Marketing, 1960–66, box 2; and U.S. Department of Transportation, National Highway Traffic Safety Administration, *Evaluation of Diagnostic Analysis and Test Equipment for Small Automotive Repair Establishments, A Report to Congress* (Washington, D.C.: GPO, 1978), 130–34.

68. Mass audience accounts of the new diagnostic centers include Devon Francis, "Here Come the Car Clinics," *Popular Science,* March 1963, 53–56, 230; "Electronic Clinic for Sick Cars," *Popular Mechanics,* January 1966, 116–19; "New Approach in Auto Repair: Mobil's Diagnostic Clinics," *Readers Digest,* January 1967, 123–26; "Automatic Doctors for Autos: Diagnostic Centers," *Business Week,* 30 August 1969, 78–81; "New Home for Solving #1 Car Problem! Diagnostic Repair Center," *Better Homes and Gardens,* November 1966, 46. Industry publicity and discussion of diagnostic centers include "Repair Centers: Is the Profit Really There?" *National Petroleum News,* June 1963, 116–24; and the following series of articles from *Motor Age* reprinted in U.S. Senate, Committee on Commerce, *Motor Vehicle Diagnostic Analysis Technology,* 723–56: "How to Equip for Diagnostic Service," March 1967; "Productivity in Diagnosis," March 1969; "Diagnosis: More Growth, Fewer Pains," February 1969; "Independents Do Their Own Thing," February 1970; "Oils Explode with Mini-Centers," February 1970; "Giant Centers Spawn Ideas, Not Dollars," February 1970; "Impact of the Diagnostic Idea," February 1970; "Diagnosis Reaches Out," February 1971. *Motor Age 1970 Census of Diagnostic Centers,* rpt. in U.S. Senate, Committee on Commerce, *Motor Vehicle Diagnostic Analysis Technology,* 686–722. On J. C. Penney's centers, see conference address of Wilbert B. Dubin, Automotive Merchandising and Service manager, J. C. Penney Co., in U.S. Senate, Committee on Commerce, *Motor Vehicle Diagnostic Analysis Technology,* 42–60. On Ford's center see press release, 31 January 1967, floor plan, and artist rendering, FMC Archives, AR-70-20, Executive Files: J. S. Andrews, President, Ford of Europe, European Automotive Group, 1946–61, 1964–66, 1970, box 2.

69. U.S. Senate, Committee on Commerce, *Motor Vehicle Diagnostic Analysis Technology,* 62–65, 68.

70. Charles B. Camp, "The Owners of Clinics That Diagnose Car Ills Have Some Headaches," *Wall Street Journal,* 17 October, 1968; "Diagnostic Centers," *Washington Consumers' Checkbook* 1, no. 2, Summer 1976, 86–92.

71. Testimonies of Harry Wright and James W. Hall, *Automotive Repair Industry Hearings*, pt. 1, 304–37.

72. The most well-known test-only center was the Auto Club of Missouri's center in St. Louis. In addition to this center, and Kreigel's, the *Motor Age 1970 Census of Diagnostic Centers* listed Auto Analysts in Sacramento, the California State Automobile Association in San Francisco, and the Riverside [Calif.] Auto Lab as the only other retail test-only diagnostic centers.

73. Donald Randall and Arthur P. Glickman, *The Great American Auto Repair Robbery: A Report on a Ten-Billion Dollar National Swindle and What You Can Do about It* (New York: Charterhouse, 1972), 155.

74. U.S. Senate, Committee on Commerce, *Motor Vehicle Diagnostic Analysis Technology*, 9.

75. U.S. Statutes at Large 86 (1972): 959–61.

76. U.S. Department of Transportation, National Highway Traffic Safety Administration, *Evaluation of Diagnostic Analysis and Test Equipment of Small Automotive Repair Establishments* (Washington, D.C.: GPO, 1978), 2, 9.

77. Devon Francis, "Smokey Matches Wits with a Computerized Car Clinic," *Popular Science*, December 1968, 74–77, 177.

78. U.S. Senate, Committee on Commerce, *Motor Vehicle Diagnostic Analysis Technology*, 120, 792.

79. Ibid., 82–116.

80. See photograph of army mechanic performing test drive while wearing diagnostic unit headphones, U.S. Senate, Committee on Commerce, *Motor Vehicle Diagnostic Analysis Technology*, 103.

81. U.S. Senate, Committee on Commerce, *Motor Vehicle Diagnostic Analysis Technology*, 72–73.

82. Josef Metz, vice president of Corporate Service for Volkswagen of America, claimed his company's plug and diagnostic system "eliminates the possibility of human error." U.S. Senate, Committee on Commerce, *Motor Vehicle Diagnostic Analysis Technology*, 15–19.

83. Donald Randall relayed this rumor in House Committee on Interstate and Foreign Commerce, Subcommittee on Consumer Protection and Finance, *Motor Vehicle Information and Cost Savings Act of 1972—Oversight*, 95th Cong., 1st sess., 1977, 144.

84. House Committee on Interstate and Foreign Commerce, Subcommittee on Consumer Protection and Finance, *Auto Repair*, 95th Cong., 2nd sess., 14, 20, 21, 25 September, 19 October, and 4 December 1978.

85. House Committee, *Auto Repair*, 5–21.

86. Doyle, *Taken for a Ride*, 61–63; Christopher J. Bailey, *Congress and Air Pollution: Environmental Policies in the USA* (Manchester: Manchester University Press, 1998), 125–40.

87. Tetraethyl lead, an octane-boosting additive in gasoline also fouled catalytic converters. The long-known health hazards of lead contamination did not move oil refiners to eliminate lead additives, but a 1970 visit from General Motors president Ed Cole to a meeting of the American Petroleum Institute (API) started the ball rolling. Cole announced GM's intention to use catalytic converters to meet emissions requirements, and API members' use of lead stood in the way. The U.S. Congress eventually stepped in and

mandated unleaded fuel be made available nationwide in 1972 and that leaded gasoline be phased out entirely by 1997.

88. See interview with Stork in Joseph M. Callahan, "Tightening the Emissions Screws," *Automotive Industries*, 15 September 1977, 23–26.

89. U.S. Environmental Protection Agency, Office of Mobile Source Air Pollution Control, Advisory Circular no. 91, 8 February 1980.

90. EGR-related recall numbers derived from compilation of emissions recalls in Doyle, *Taken for a Ride*, 515–32.

91. For details on emissions-related technologies, see J. Robert Mondt, *Cleaner Cars: The History and Technology of Emission Control since the 1960s* (Warrendale, Pa.: SAE, 2000); Ronald K. Jurgen, ed., *History of Automotive Electronics*, 3 vols. (Warrendale, Pa.: SAE, 1998); and Jurgen, *On- and Off-Board Diagnostics* (Warrendale, Pa.: SAE, 2000).

92. U.S. Senate, Committee on Commerce, *Motor Vehicle Diagnostic Analysis Technology*, 118–19.

93. Knudsen, *Vehicle Emission Measurement*, 17–18.

94. "Statement of Frank A. Krich, Regulatory Planning Specialist, Environmental Energy Affairs, Chrysler Corp.," U.S. Senate, Committee on Commerce, Science, and Transportation, *Auto Repair Fraud: Hearing before the Subcommittee on Consumer*, 102nd Cong., 2nd sess., 21 July 1992, 62–63. See also Julie Edelson Halpert, "Who Will Fix Tomorrow's Cars?: A Battle Is Raging over the Keys to the Coming Computerized Repair Systems," *New York Times*, 7 November 1993; Dale Jewett, "EPA Referees Chip Reprogramming," *Automotive News*, December 1993.

95. Quote from Ford advertisement in *Automotive News, Insight Supplement*, 8 July 1991, 3i.

96. "You'll Wonder How You Got Along without It," advertisement, *Motor Age*, September 1998, 8–9.

97. Dale Jewett, "Quick Fix," *Automotive News, Insight Supplement*, 8 July 1991, 20i.

98. "Say, Smokey—," *Popular Science*, May 1985, 42.

99. See the various articles on the topic of technical work in the post–World War II era in Stephen R. Barley and Julian E. Orr, eds., *Between Craft and Science: Technical Work in U.S. Settings* (Ithaca, N.Y.: Cornell University Press, 1997); as well as the influential study of computer automation in settings such as paper making and clerical work in Shoshana Zuboff, *The Age of the Smart Machine: The Future of Work and Power* (New York: Basic Books, 1988).

100. Smith, qtd. in Ann Dykman, "Smart Cars Need Smart Techs," *Techniques*, 72, October 1997, 36–38.

101. Jennifer Bott, "The Proud, the Few: Auto Mechanics are in High Demand," *Detroit Free Press*, 12 November 1998. See also Gabriella Stern, "Less Tinkering, More Computing," *Detroit Free Press*, 28 August 1997; Warren Brown, "Car Dealers Cite Shortage of Qualified Technicians," *Washington Post*, 12 February 1995; Warren Brown, "Wrenching Frustrations: Autos' High-Tech Features, Shortage of Good Mechanics Make Repairing the Family Car Tougher and Costlier," *Washington Post*, 2 December 1995. See also Kathleen McGraw and Robert Forrant, *Workers' Perspective: Skills, Training, and Education in the*

Automotive Repair, Printing, and Metalworking Trades (Berkeley, Calif.: National Center for Research in Vocational Education, 1992).

Conclusion • Servants or Savants?

1. Frank R. Wilson, *The Hand: How Its Use Shapes the Brain, Language, and Human Culture* (New York: Pantheon Books, 1998), 274–75.

2. Others have explored gender and racial hierarchies in traditional technological domains such as the engineering profession or the factory floor and in what were previously overlooked technological domains such as the home or the classroom. Recent studies have even focused on such hierarchies in automobile marketing, imagery, and usage. Ruth Oldenziel, *Making Technology Masculine: Men, Women, and Modern Machines in America, 1870–1945* (Amsterdam: Amsterdam University Press, 1999); Cynthia Cockburn, *Machinery of Dominance: Women, Men, and Technical Know-How* (London: Pluto Press, 1985); Ruth Schwartz Cowan, *More Work for Mother: The Ironies of Household Technology from the Open Hearth to the Microwave* (New York: Basic Books, 1983); Nina E. Lerman, "'Preparing for the Duties and Practical Business of Life': Technological Knowledge and Social Structure in Mid-19th Century Philadelphia," *Technology and Culture* 38 (January 1997): 31–59; Bruce Sinclair, ed., *Technology and the African-American Experience: Needs and Opportunities for Study* (Cambridge, Mass.: MIT Press, 2004); Julie Wosk, *Women and the Machine: Representations from the Spinning Wheel to the Electronic Age* (Baltimore: Johns Hopkins University Press, 2001); Kathleen Franz, *Tinkering: Consumers Reinvent the Early Automobile* (Philadelphia: University of Pennsylvania Press, 2005)

3. For further elaboration of the theoretical insights to be gained by studying the chauffeur problem, see Kevin Borg, "The 'Chauffeur Problem' in the Early Auto Era: Structuration Theory and the Users of Technology," *Technology and Culture* 40 (October 1999): 797–832.

4. I have borrowed this insightful phrase from Mike Rose's study "'Our Hands Will Know': The Development of Tactile Diagnostic Skill—Teaching, Learning, and Situated Cognition in a Physical Therapy Program," *Anthropology and Education Quarterly* 30, June 1999, 133–60.

5. A production-consumption view, for example, explains why automakers tried to increase sales of new cars to customers by improving reliability and serviceability through design. This explains for some the decision by Ford to use vanadium steel in Model T axles: they would last longer, cause fewer service problems, improve the company's reputation among consumers, and ultimately lead to increased sales of new cars.

6. Ruth Schwartz Cowan inspired what has now been nearly two decades of fruitful research in the history of technology by calling on sociologists and historians of technology to shift attention from invention and production to the "consumption junction," where consumers made the decisions about which technologies succeeded and which failed. "All technologies have consumers," she wrote, "and all technological development is oriented toward a positive consumption decision, whether the ultimate consumers are located in the consumption domain (as householders are) or in some other domain (wholesale, retail,

or production)." Ruth Schwartz Cowan, "The Consumption Junction: A Proposal for Research Strategies in the Sociology of Technology," in *The Social Construction of Technological Systems,* ed. Wiebe E. Bijker, Thomas P. Hughes and Trevor Pinch (Cambridge, Mass.: MIT Press, 1987), 261–80.

7. George Basalla's model of the evolution of technology by the cultural selection from many possible technical variations is relevant here. *The Evolution of Technology* (Cambridge: Cambridge University Press, 1988).

8. Thomas Misa calls for a "meso level" of analysis, conceptually intermediate between the macro and the micro, which might be achieved by "analyzing the institutions intermediate between the firm and the market or between the individual and the state." Misa does not mention the repair shop as a candidate site, but this study arguably addresses the dilemma he and others have described. Reinhold Reith argues, as I do, that the study of repair can shed new light on everyday sociotechnical phenomena in ways that studies of production and consumption have not. Thomas J. Misa, "Retrieving Sociotechnical Change from Technological Determinism," in *Does Technology Drive History? The Dilemma of Technological Determinism,* ed. Merritt Roe Smith and Leo Marx (Cambridge, Mass.: MIT Press, 1994), 115–41; Reinhold Reith, "Reparien: Ein Thema der Technikgeschichte?" in *Kleine Betriebe—angepaßte Technologie? Hoffnungen, Erfahrungen und Ernüchterungen aus sozial-und technikhistorischer Sicht,* ed. Reinhold Reith and Dorothea Schmidt (Münster: Waxmann, 2002), 139–63.

9. Bob Weber, "Going the Distance for Learning," *Motor Age* 116, February 1997, 18–25.

10. Peter Whalley and Stephen R. Barley, "Technical Work in the Division of Labor: Stalking the Wily Anomaly," in Barley and Orr, *Between Craft and Science: Technical Work in U.S. Settings,* 39.

11. See Mike Rose, *The Mind at Work: Valuing the Intelligence of the American Worker* (New York: Viking, 2004); Charles M. Keller and Janet Dixon Keller, *Cognition and Tool Use: The Blacksmith at Work* (Cambridge: Cambridge University Press, 1996); Julian Orr, *Talking about Machines: An Ethnography of a Modern Job* (Ithaca, N.Y.: Cornell University Press, 1996).

12. Wilson, *Hand;* Jean Lave and Etienne Wenger, *Situated Learning: Legitimate Peripheral Participation* (Cambridge: Cambridge University Press, 1991).

13. Shoshana Zuboff, *The Age of the Smart Machine: The Future of Work and Power* (New York: Basic Books, 1988), 392–93.

14. Richard Sennett and Jonathan Cobb, *The Hidden Injuries of Class* (New York: W. W. Norton Co., 1972), 215–16.

15. This is a growing concern among many education researchers, and members of the automotive service industry would do well to participate and contribute to this reassessment of educational dichotomies.

16. California State Legislature, *Public Hearing on Certification of Automotive Mechanics,* Los Angeles, 26 November 1974, 48–49.

ESSAY ON SOURCES

When it comes to historical sources, auto repair is everywhere and nowhere. Auto mechanics and repair shops existed in virtually every community across America, but there remains no conventional cache of business or labor records from which to work. The occupation's social status had repercussions here as well. Auto mechanics and repair shops did not often keep sufficient business or personal records. Even if they had, few archives or libraries sought them out. Labor organizations such as the International Association of Machinists gained only a tiny foothold in the industry, so their records shed limited light on mechanics' experience. Therefore, researchers studying this topic must tap a wide range of sources. I will attempt here to draw the interested researcher's attention to some of the sources I found fruitful as well as a few promising ones that I had to leave untapped. Additional sources and greater detail are provided in the chapter notes.

Scholarship on the automobile manufacturing industry abounds and must be mastered prior to delving into the automobile repair industry. Thankfully, Michael Berger has made this task much easier with publication of *The Automobile in American History and Culture: A Reference Guide* (Westport, Conn.: Greenwood Press, 2001). Berger's masterful bibliographic guide makes a reiteration of the major automotive history scholarship here unnecessary. He does not highlight scholarly studies focused on the automobile service industry, however, because they are so few. James J. Flink's classic work *America Adopts the Automobile, 1895–1910* (Cambridge, Mass.: MIT Press, 1970) devoted one subchapter to the subject of maintenance and repair as well as one to "mechanical expertise." Henry Dominguez published numerous photographs of early Ford agencies in *The Ford Agency: A Pictorial History* (Osceola, Wis.: Motorbooks International, 1980). Douglas Harper, *Working Knowledge: Skill and Community in a Small Shop* (Chicago: University of Chicago Press, 1987), while not a historical study, presents an insightful ethnography of a contemporary automobile mechanic that highlights important and enduring qualities of mechanics' knowledge and social relationships. Thomas S. Dicke, *Franchising in America: The Development of a Business Method, 1840–1980* (Chapel Hill: University of North Carolina Press, 1992), contains chapters on the Ford Motor Co. and Sun Oil Co. that touch on service from the perspective of marketing new products. John A. Jakle and Keith A. Sculle, *The Gas Station in America* (Baltimore: Johns Hopkins University Press, 1994); and John Margolies, *Pump and Circumstance: Glory Days of the Gas Station* (Boston: Little, Brown, 1993), both touch on, but do not delve into, the repair services offered by gas stations.

Joseph Corn drew attention specifically to auto repair in his 1992 article, "Work and Vehicles: A Comment and a Note," in *The Car and the City: The Automobile, the Built Environment, and Daily Urban Life,* ed. Martin Wachs and Margaret Crawford (Ann Arbor: University of Michigan Press, 1992), 25–34. Corn rightly suggested that scholarly historical attention to the work done on automobiles might prove insightful and fruitful because the relationship between motorists and mechanics has long been "one of the most psychologically charged relationships of modern consumer societies." Finally, in 1995 Stephen McIntyre completed the first sustained scholarly study of the topic in " 'The Repairman Will Gyp You': Mechanics, Managers, and Customers in the Automobile Repair Industry, 1896–1940" (Ph.D. diss. University of Missouri, Columbia, 1995). McIntyre staked out the contours of the auto repair industry over the first half of the twentieth century and focused a labor historian's eye on the perennial discontent motorists felt with the automobile service industry, locating its source in class-based conflicts, both between mechanics and managers and between motorists and mechanics. McIntyre's research, a portion of which appeared as "The Failure of Fordism: Reform in the Automobile Repair Industry, 1913–1940," *Technology and Culture* 41 (April 2000): 269–99, proved foundational for the present study and should be consulted by anyone embarking on further research in this area. Valuable nonhistorical studies of auto mechanics in later years include Jim L. Drost, "Job Characteristics of Automotive Mechanics in Selected Iowa Dealerships and Garages" (Ph.D. diss., Iowa State University, Ames, 1970); and Bonalyn J. Nelson, "The Nature and Implications of Technological Change and the Rise of a Service Economy: Observations from the Field of Automotive Repair" (Ph.D. diss., Cornell University, 1998).

The Henry Ford Museum and Greenfield Village, Dearborn, Mich., holds a range of primary source material reflecting that automaker's perspective on the automobile service industry as manifest in dealership service departments. Particularly useful are the General Letters, Branch Letters, Service Dept. Letters, Sales Literature, and Auto Parts and Service collections. The Ford Motor Company Archives, formerly the Ford Industrial Archives, hold additional relevant material not transferred to the Henry Ford Museum, including the Executive Files of L. A. Iacocca, 1967; E. R. Laux, 1960–67; Wright Tinsdale, 1957–72; and J. S. Andrews, 1946–70; as well as the publications *Service Research Center Quarterly* and *Ford Service Life*. The Smithsonian Institution's National Museum of American History (NMAH) holds a very useful Transportation Collection, including the partial records of a Washington, D.C., repair business Call Carl's, one of the few surviving record sets of an independent repair shop. The NMAH's Trade Catalog Collection and the Warshaw Collection in its Archives Center also contain various printed matter related to automobile parts, tools, equipment, and schools. Similar printed materials can be found in the Hagley Museum and Library Trade Catalog collection, Wilmington, Del.; the Romaine Trade Catalog Collection, University of California, Santa Barbara; the Philadelphia Free Library's Automobile Reference Collection; and the Detroit Public Library's Automotive History Collection. One can also find the A. L. Dyke Collection at the Detroit Public Library collection, with some of the personal and business records of one of the most prolific writers of technical literature for mechanics and one of the pioneers of correspondence education for those seeking to enter the occupation. Particularly promising are the numerous testimonial letters from customers of his books and correspondence course. These letters were not processed at the time of my visit to the library, and I did not have time to delve deeply into them. They await an eager graduate student. One of Dyke's home study aids with stamped metal moving parts is preserved in the Transportation Collection at the NMAH.

A relative wealth of primary source records exists relating to the training and education of automobile mechanics. I have found the archives of the YMCA of Greater New York very helpful for understanding the extent of the chapter's educational activities. I pestered the New York YMCA organization for two years at the outset of my research and finally gained access. I am not certain if they remain available to researchers. The YMCA of the USA Archives at the University of Minnesota provided me with copies of the limited amount of mechanic training material available in its collection. Particularly intriguing were images of YMCA auto courses offered overseas. The Boston YMCA also conducted an early Automobile School, and its records remain largely untapped at Northeastern University's Snell Library. The Knights of Columbus Supreme Council Archives in New Haven, Conn., hold some material relating to the automotive courses the organization offered from the conclusion of World War I into the late 1920s. The U.S. Army Military History Institute in Carlisle, Pa., holds a small collection of World War I veteran surveys which shed light on the training and deployment of mechanics in the Motor Transportation Corps, including David McNeal's diary, which proved particularly useful for this study. The National Archives in Washington, D.C., holds additional unit histories and administrative records from World War I. The Records of the War Department General and Special Staffs, Committee on Education and Special Training, are especially useful. Studying John C. Speedy III, "From Mules to Motors: Development of Maintenance Doctrine for Motor Vehicles by the U.S. Army, 1896–1918" (Ph.D. diss., Duke University, 1977), will make research in the Military History Institute or the National Archives much more productive.

The records of the U.S. Office of education at the National Archives, College Park, Md., provide a window into the institutionalization of vocational auto shop in the public school systems across the United States. Particularly valuable are the Records of the Assistant Commissioner for Vocational Education, which contain annual state reports of numbers of courses offered, locations, enrollments, teachers employed, funding, and more from 1917 through 1937. See also the Papers of John C. Wright, assistant commissioner for Vocational Education for the U.S. Office of Education, at the Strozier Library, Florida State University, Tallahassee. Journals such as *Industrial Arts Magazine, Industrial Arts and Vocational Education,* and *Manual Training Magazine* captured educators' views on creating mechanics, curriculum outlines, and lesson plans. A number of education master's theses and university and school district studies also provide detailed pictures of particular auto shop programs. See Harry W. Paine, "A Survey of the Boys' Technical High School of Milwaukee, Wisconsin, and the Organization of Its Automotive Department" (master's thesis, Iowa State College, 1928); Melvin S. Lewis, *Analysis of the Automechanic's Trade with Job Instruction Sheets,* Division of Vocational Education of the University of California and the State Board of Education, Trade and Industrial Series No. 4 (Berkeley: University of California, 1925); Oakland Public Schools, *Automobile Repair in the Vocational Continuation School,* Superintendent's Bulletin No. 33, Course of Study Series (Oakland, Calif.: Oakland Public Schools, 1922); Lewis S. Neeb, "The Automobile as a Subject of Instruction in the Public Secondary Schools" (master's thesis, University of Arizona, 1927); and Lynn C. McKee, "A Trade Training Curriculum in Automobile Mechanics for Senior High Schools" (master's thesis, Duke University, 1931).

The broader historical development of vocational education in the United States has been the subject of numerous studies. A good place to begin is with

Paul Willis's classic study of how the British school system often perpetuated class divisions in *Learning to Labor: How Working Class Kids Get Working Class Jobs* (New York: Columbia University Press, 1977). On U.S. schools, see Paul C. Violas, *The Training of the Urban Working Class: A History of Twentieth Century American Education* (Chicago: Rand McNally, 1978); Ira Katznelson and Margaret Weir, *Schooling for All: Class, Race, and the Decline of the Democratic Ideal* (New York: Basic Books, 1985); Marvin Lazerson and W. Norton Grubb, eds., *American Education and Vocationalism: A Documentary History, 1870–1970*, Classics in Education, no. 48 (New York: Teachers College Press, 1974); Arthur F. McClure, James Riley Chrisman, and Perry Mock, *Education for Work: The Historical Evolution of Vocational and Distributive Education in America* (London: Associated University Presses, 1985); Clyde W. Hall, *Black Vocational, Technical and Industrial Arts Education: Development and History* (Chicago: American Technical Society, 1973); Ambrose Caliver, *Vocational Education and Guidance of Negroes: Report of a Survey Conducted at the Office of Education* (1937; rpt., Westport, Conn.: Negro Universities Press, 1970); and Nina E. Lerman, "From 'Useful Knowledge' to 'Habits of Industry': Gender, Race, and Class in Nineteenth-Century Technical Education" (Ph.D. diss., University of Pennsylvania, 1993).

Various serial publications contain material reflecting developments in the automobile service industry. Primary among the trade journals cited in this study are the *Automobile, Automobile Dealer and Repairer, Automobile Topics, Automobile Trade Journal, Automotive Engineering, Automotive Industries, Automotive Merchandising, Engineering News-Record, Society of Automotive Engineers Journal , Horseless Age, Jobber Topics, Motor Age,* and *Motor World.* Journals of related trades that bear on the auto repair industry include *American Machinist, American Blacksmith, Bicycling World, Blacksmith and Wheelwright,* the *Chauffeur,* the *Garage,* and the *Hub.* Popular press accounts can be found in a wide range of periodicals, including *Detroit Free Press, Illustrated World, Mechanix Illustrated, New York Times, Outing Magazine, Philadelphia Free Press, Philadelphia Inquirer, Popular Mechanics, Popular Science, Reader's Digest, Scientific American,* and *Washington Post.* Many more can be traced through the *Readers' Guide to Periodical Literature.*

As K. Austin Kerr, Amos J. Loveday, and Mansel G. Blackford note in, *Local Businesses: Exploring Their History* (Nashville, Tenn.: American Association for State and Local History, 1990), local libraries and historical societies can prove useful sources for recovering the history of small businesses. The same is true for auto mechanics and their world. The unassuming Feldheim Library in San Bernardino, Calif., and the downtown library in Riverside, Calif., for example, yielded city directories, maps, clipping files, high school yearbooks, and other material useful for studying the ubiquitous nature of the automobile repair industry. While blacksmith's records from the early twentieth century are less common in research collections compared to those from the nineteenth century, the John Vander Voort daybooks at Rutgers University Library record the arrival of automobiles in a local tradesman's shop. No doubt other examples could be unearthed in local libraries and historical societies' collections. This study, of necessity, did not delve into regional variations in the industry, but these community resources would be the place to start such a follow-up inquiry.

African Americans' experiences in the automobile repair industry are reflected in much of the material already mentioned but are examined more explicitly in Virginia Peeler, *The Colored Garage Worker in New Orleans* (New Orleans: High School Scholarship Association,, 1929); J. H. Harmon Jr., "The Negro as a Local Business Man," *Journal of Negro History* 14 (April 1929): 116–55; U.S. Department of Commerce, Bureau of the Census, *Negroes in the United States, 1920–32* (Washington, D.C.: GPO, 1934), chap. 17: "Retail Business"; and Bernard Levenson and Mary S. McDill, "Vocational Graduates in Auto Mechanics: A Follow-up Study of Negro and White Youth," *Phylon: The Atlanta University Review of Race and Culture* 27 (Winter 1966): 347–57. A sense of the vast number of black-owned automobile businesses in communities across the United States can be gained by surveying the advertisements in publications such as early volumes of W.E.B. DuBois's the *Crisis;* James N. Simms, comp., *Simms' Blue Book and National Negro Business and Professional Directory* (Chicago: James N. Simms, 1923); R. Irving Boone, ed., *Negro Business and Professional Men and Women: A Survey of Negro Progress in Varied Sections of North Carolina,* vol. 2: *1946* (Wilmington, N.C.: R. Irving Boone, 1946); and the entries in the various black motoring guides published from 1936 through the mid-1960s, including *The Baker Handbook, GO Guide to Pleasant Motoring, The Negro Motorist Green-book,* and *Travelguide.* The widely available microform copy of the Schomburg Center Clipping File also includes directories, advertisements, newspaper articles, and surveys such as the National Negro Business League, "Report of the Survey of Negro Business." Such business surveys often include automotive businesses, though scholars of black businesses have rarely studied them. This absence is explained in part in Robert E. Weems Jr.,·"Out of the Shadows: Business Enterprise and African American Historiography," *Business and Economic History* 26 (Fall 1997): 200–212. The few related historical studies include August Meier and Elliott Rudwick, "The Boycott Movement against Jim Crow Streetcars in the South, 1900–1906," *Journal of American History* 55 (March 1969): 756–75; Blaine A. Brownell, "A Symbol of Modernity: Attitudes toward the Automobile in Southern Cities in the 1920s,"

American Quarterly 24 (March 1972): 20–44; Kathleen Franz, "'The Open Road': Automobility and Racial Uplift in the Interwar Years," in *Technology and the African-American Experience: Needs and Opportunities for Study,* ed. Bruce Sinclair (Cambridge, Mass.: MIT Press, 2004), 131–53.

Following World War II, increasing consumer dependence upon automobiles brought increased public attention on the auto repair industry. Roger Riis and John Patric's, *Repair Men May Gyp You,* 2nd ed. (1949), kept the "service problem" in the public dialog during the early postwar years. The best single collection of primary sources on the state of the industry by the 1960s is the four thousand–page, six-volume set of testimony and documents gathered in Senate Committee on the Judiciary, Subcommittee on Antitrust and Monopoly, *Automotive Repair Industry Hearings,* 90th Cong., 2nd sess., 6 vols., 1969–71. Generated by Philip Hart's hearings on the industry, these volumes also incorporate additional testimony on auto repair given before the Federal Trade Commission. Insightful testimony from motorists and mechanics is also available in Senate Committee on Business and Professions, California State Legislature, *Public Hearing on Certification of Automotive Mechanics,* Los Angeles, 26 November 1974. Other useful congressional investigations include, House Committee on Interstate and Foreign Commerce, Subcommittee on Consumer Protection and Finance, *Motor Vehicle Information and Cost Savings Act of 1972—Oversight,* 95th Cong., 1st sess., 2 and 9 May 1977; and House Committee on Interstate and Foreign Commerce, Subcommittee on Consumer Protection and Finance, *Auto Repair,* 95th Cong., 2nd sess., 14, 20, 21, and 25 September, 19 October and 4 December 1978. Later investigations that could support further studies into the problems of diagnostic codes, tampering, and access to and ownership of data include Senate Committee on Commerce, Science, and Transportation, Subcommittee on the Consumer, *Auto Repair Fraud,* 102nd Cong., 2nd sess., 21 July 1992, and 103rd Cong., 1st sess., 4 March 1993; and Senate Committee on Commerce, Science, and Transportation, Subcommittee on Consumer Affairs, Foreign Commerce and Tourism, *Customer Choice in Automotive Repair Shops,* 107th Cong., 2nd sess., 30 July 2002.

Sources documenting the shift in federal and state policies toward diagnostic technologies include U.S. Department of Health Education and Welfare, National Air Pollution Control Administration, *Control Techniques for Carbon Monoxide, Nitrogen Oxide, and Hydrocarbon Emissions from Mobile Sources* (Washington, D.C.: GPO, 1970); Roy F. Knudsen, ed., *Vehicle Emission Measurement—Panel Discussion: Transcription of the Vehicle Emission Measurement Panel Presented at the Instrument Society of America Conference, October 26–29, 1970* (Pittsburgh: Instrument Society of America, 1971); U.S. Senate, Committee on Commerce, *Motor Vehicle Diagnostic Analysis Technology, 1971–85,* technical conference proceedings for the use of the committee, 92nd Cong., 2nd sess., 1971, Committee Print, 43; U.S. Department of Transportation, National Highway Traffic Safety Administration, *Evaluation of Diagnostic Analysis and Test Equipment for Small Automotive Repair Establishments: A Report to Congress,* (Washington, D.C.: GPO, 1978); Burton L. Jones, Joseph F. Peters, and Bernard J. Schroer, *Selective Survey of the Capability of Representative Automobile Repair Facilities to Diagnose and Repair Automobiles,* a report prepared for the Department of Transportation, Consumer Affairs Division, May 1979; and Booz, Allen and Hamilton, Inc., *Scope and Impact of New Automotive Technology on the Inspection, Diagnosis and Repair Process,* report prepared for the U.S. Department of Transportation, National Highway Traffic Safety Administration, November 1980.

The standard historical study of California's postwar struggle with automobile tailpipe emissions is James Krier and Edmund Ursin, *Pollution and Policy: A Case Essay on California and Federal Experience with Motor Vehicle Air Pollution, 1940–1975* (Berkeley: University of California Press, 1977). Subsequent studies of the confluence of science, technology, and policy include Lawrence J. White, *The Regulation of Air Pollutant Emissions from Motor Vehicles* (Washington, D.C.: American Enterprise Institute, 1982); Christopher J. Bailey, *Congress and Air Pollution: Environmental Policies in the USA* (Manchester: Manchester University Press, 1998); and Scott Hamilton Dewey, *Don't Breathe the Air: Air Pollution and U.S. Environmental Politics, 1945–1970* (College Station: Texas A&M University Press, 2000). Jack Doyle, *Taken for a Ride: Detroit's Big Three and the Politics of Pollution* (New York: Four Walls Eight Windows, 2000), provides extensive documentation of the legislative struggles over auto emissions into the 1990s from the perspective of a former policy analyst. J. Robert Mondt provides a GM engineer's view of the resulting under-hood developments in *Cleaner Cars: The History and Technology of Emission Control since the 1960s* (Warrendale, Pa.: Society of Automotive Engineers [SAE], 2000). To follow the development of engineers' ideas about electronic and diagnostic technology, see the collections of SAE technical papers in, Ronald K. Jurgen, ed., *History of Automotive Electronics: The Early Years* (Warrendale, Pa.: SAE, 1998); and *On-and Off-Board Diagnostics,* Automotive Electronics series (Warrendale, Pa.: SAE, 2000).

INDEX